The Stress of Combat
The Combat of Stress

For my grandchildren

Alex, Christopher, Matthew, Rachael
and Christopher Brian

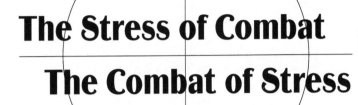

The Stress of Combat
The Combat of Stress

CARING STRATEGIES
TOWARDS EX-SERVICE
MEN AND WOMEN

ROY BROOK

Foreword by
General Sir Charles Huxtable KCB CBE

sussex
ACADEMIC
PRESS
Brighton • Portland • Toronto

Copyright © Roy Brook 1999, 2010.

The right of Roy Brook to be identified as author of this work has been asserted in accordance with the Copyright, Designs and Patents Act 1988.

2 4 6 8 10 9 7 5 3

First published 1999 by the Alpha Press in the United Kingdom, reprinted by Sussex Academic Press 2010, with additional chapter ("Bringing the Story up to Date") and revised and updated Appendix C: Useful Addresses – Where to go to for Help.
SUSSEX ACADEMIC PRESS
PO Box 139
Eastbourne BN24 9BP

and in the United States of America by
SUSSEX ACADEMIC PRESS
920 NE 58th Ave   Suite 300
Portland, Oregon 97213–3786

and in Canada by
SUSSEX ACADEMIC PRESS (CANADA)
90 Arnold Avenue, Thornhill, Ontario L4J 1B5

British Library Cataloguing in Publication Data
A CIP catalogue record for this book is available from the British Library.

Library of Congress Cataloging-in-Publication Data
Roy, Brook.
The stress of combat : the combat of stress / Roy Brook.
p. cm.
Includes bibliographical references and index.
  ISBN 978-1-84519-407-9 (original Alpha Press ISBN, 1-898595-27-5)
(pbk. : alk. paper)
1. Veterans—Mental health—Great Britain. 2. Veterans—Mental health services—Great Britain. 3. Post-traumatic stress disorder—Great Britain. I. Title.
RC451.4.V48 B76 1999
616.85'212'00941—dc21

                                                         98-46657

Printed by TJ International, Padstow, Cornwall.
Printed on acid-free paper.

Contents

Foreword

By Sir Charles Huxtable KCB CBE

Before I became President of The Ex-Services Mental Welfare Society, Combat Stress, in 1991, I had been aware of the Society and its work, but I had little understanding of the scope and extent of the task facing Combat Stress. I found that, whilst those who were involved with the Society in any way were very supportive and well informed, the public at large were generally unaware of the problems of post traumatic stress disorder (PTSD), and the work of Combat Stress. Since then, with the wars in Iraq and Afghanistan, public awareness has grown enormously. This is partly as a result of the additional publicity given to PTSD in Parliament and in the Press, and partly as a result of the efforts of Roy Brook with his first edition of this book.

It is fortunate that awareness has grown as the problem of PTSD is not only still with us but is increasing at an alarming rate. Demand for the Society's services has risen by 53% over the past three years. The average time between a Serviceman or woman being discharged and seeking the Society's help is 13½ years. The Society is still acquiring clients from the Falklands war and as a result of service in Northern Ireland as well as the Balkans. Whilst there is already a growing number of veterans of the Iraq and Afghanistan wars making contact with the Society, there is no doubt that the number seeking the Society's help as a result of their experiences in those wars will continue to grow for many years.

It is therefore very important that the nation should understand the problem and the need to ensure that proper help and support is provided for those whose peace of mind has suffered as a result of their efforts in the service of their country. This is why I am pleased to be able to welcome the publication of the updated edition of "The Stress of Combat, The Combat of Stress".

Charles Huxtable

Educated at Wellington College, Charles Huxtable was commissioned into the Duke of Wellington's Regiment in February 1952 and served as a platoon commander in the latter stages of the Korean War. In 1967 he served as a Company Commander in Cyprus. In 1980 he was appointed Commander of Land Forces in Northern Ireland. He was Director of Army Staff Duties between 1982 and 1983 and then Commander of Training and Arms Directors at the Ministry of Defence from 1983 to 1986. He was Quartermaster-General from 1986 to 1988. He served as the Commander in Chief, UK Land Forces from 1988 to 1990 when he retired.

Preface

This book is dedicated to the hundreds of mentally disabled ex-Servicemen and women I have had the honour to serve over the past ten years as a welfare officer with one of the national ex-Service charities.

Some, alas, are no longer alive. They range from First World War veterans in their nineties to young men who served in the Gulf or Northern Ireland; from senior officers to raw recruits. Some were reluctant to talk about their experiences. Others could not help recounting their war stories which dominated every day of their lives. Some of these individuals were well supported by a caring wife or husband, while others had been abandoned long ago to exist in very poor circumstances.

Hardly a week passed without news of new clients, many of whom I had never heard of before. While we simply do not know the extent of the problem nationally, we do know that there are significant numbers of men and women who have served their country well and who are in serious need of social and financial support.

This book is a tribute to these men and women. It is written with a view to informing and alerting the general public, the Service and ex-Service authorities, and the caring professions of the problem. There are *thousands of cases* where an individual has served his or her country, often in situations of great danger, but who now needs our support and understanding if we are to improve the remainder of their lives.

The Ex-Services Mental Welfare Society, founded shortly after the First World War, has always aimed to help restore ex-Service men and women to health in mind and body, particularly helping those suffering from mental illnesses. Some 50,000 cases were dealt with in the first seventy years, and currently over 3,000 veterans are being helped. However, as I intend to show in this book, there were, and still are, many not able to be helped in this way, for a whole variety of reasons. The Society continues to extend and develop its operations, but there will always be more remaining to be done. The Society deserves the support of the public at large, and of the government, as indeed do all the other societies who work in this and related fields. As social patterns change, and progress is achieved in medical science, we as a nation must recognise our responsibility to those who took up combat or served as peace-keepers. Where the *Stress of Combat* results in a mental or social condition that impacts negatively on the life of the sufferer, and his or her family, it is society's task to support those institutions and societies whose primary aim is to *Combat the Stress* of those who have suffered in serving us.

Acknowledgements

The author and publisher would like to thank Michael Haslam and Butterworth Heinemann for permission to reproduce the chart on p. 215, which was originally published in *Psychiatry Made Simple*, 1993.

When I first joined the Society, in 1984, our Chief Consultant Psychiatrist was Brigadier (Retd.) Desmond Murphy FRCPsych., late of the Royal Army Medical Corps. I was extremely fortunate in being able to sit in with Dr Murphy as he interviewed my patients when they visited Tyrwhitt House. I also sat in with him at Head Office when he was considering cases for admission to the Society's Homes. I was able to ask questions afterwards and through him I was to acquire a lasting interest in his subject and a deep sense of compassion for my patients. All that I subsequently learned in this field has reinforced what Desmond first taught me.

For the first few years of my work, when I covered the West Midlands, I was a regular visitor to the War Pensions Office in Birmingham where Colin Armstrong was Manager. I received tremendous support from Colin and his staff. When he was, deservedly, promoted, we kept in touch. He was very supportive of the idea of my writing this book. When he died, sadly, a couple years ago, he was Southern Area Manager for the War Pensions Agency.

In the final period before I retired, I was able to establish a Southern Regional Office at Alton where I was very fortunate in being able to appoint Sheila Chadwick as my Secretary, later Office Manager. Sheila has been a tower of strength, but I know she would appreciate a mention of her husband, David, who died recently. David read the original draft of the book and it was he who, with contacts in book publishing, encouraged me to go ahead and finish the work. He would have been delighted to have seen it published.

The insignia on the back cover are those of the four ex-Service charitable organisations without whom the work described in this book would not have been possible. Without them, the author would not have been aware of many of the clients and their problems. There are, of course, many other organisations dedicated to the ex-Service cause or to the medical and welfare needs from which many ex-Service individuals suffer. A comprehensive list of all these organisations known to the author can be found in Appendix C.

I am grateful for all the information and advice received from Brigadier Tony Dixon OBE, the ex-Services Mental Welfare Society; Surgeon Captain Morgan O'Connell FRC Psych; Wing Commander Gordon Turnbull FRC Psych; John Franklin, the Royal British Legion; Mrs Beila Best OBE, SSAFA Forces Help; Steve Johnson and Alan Burnham, the War Pensions Agency; and Lieutenant Colonel James Tedder, the 'Not Forgotten' Association.

Substantial assistance has been received from the staff of Poole Central Library throughout the whole period whilst the book was being written, in Hamworthy. Finally, I must thank my wife, Vera, who travelled thousands of miles with me, meeting many of the clients and their carers, and giving support at all times.

Poole, August 1998

I must first of all thank my friend, the former Chairman of the Society, General Sir Charles Huxtable KCB CBE, for kindly contributing a Foreword to this updated Edition.

Sadly, my wife, Vera, died some years ago but I was later able to marry Hazel whose late husband was a Royal Air Force Veteran. She has given me every support and we are indebted to our daughter, Barbara, for bringing the Appendices up to date via the Internet.

Poole, February 2010

Introduction

In 1984 I responded to an advertisement through the Officers' Association of the Royal British Legion. The appointment on offer was for an ex-regular officer to visit and befriend mentally disabled ex-Servicemen and women from all three Services and the Merchant Navy. As I had twenty years of military Service behind me, at home and overseas, I felt qualified to take up the task.

Over the next ten years I covered an area which ran from the West Midlands and South Wales to London, and the whole of the South Coast from Kent to Cornwall. Gradually, others were appointed to take over various parts of this large territory. I motored about a quarter of a million miles and made about five thousand visits to clients in their own homes. Although the majority of my cases were from the Second World War, every conflict since that time has been represented. Then there were the victims of both Service and civilian accidents, and also a number of mental illnesses not related to military Service at all. Some suffered from infirmities usually associated with the elderly, such as Alzheimer's disease; other conditions, such as epilepsy, were encountered in all ages.

Post Traumatic Stress Disorder (PTSD) is today the most commonly encountered mental disorder amongst Servicemen, and civilians, who have encountered a very serious and sudden shock to their system. Prior to the war in Vietnam, when PTSD was first defined as a condition in itself, thousands of sufferers from PTSD had been diagnosed as depressive, schizophrenic, or 'psychotic' in general terms. The important factor of PTSD is that it is a *post-traumatic* disorder which can occur and re-occur many years after the original incident or stressful situation. A lack of understanding of the plight of many clients has led to the break-up of marriages, the abandonment of parents by their children, and vice versa, and sometimes even suicide. Help for clients takes the form of respite care, help towards a War Disability Pension; or help from the local authority, from the GP, or from the other ex-Service organisations.

Many of the cases were lonely people, sometimes living in appalling conditions caused by neglect or the inability of the client to look after himself.[1] For many, it is enough merely to meet someone who really cares about them. The Ex-Services Mental Welfare Society offers temporary respite care in three convalescent homes, and there is also a permanent home for veterans. The benefits and limitations of these facilities will be discussed and explored. Mental and physical disabilities can be quite unrelated although sometimes the one can be brought on or aggravated by the other.

No one knows the extent of the problems to be discussed here. All I can say with any degree of certainty is that at present, clients who die are greatly outnumbered by new cases which come to light from war pensions' applications, referrals from GPs, social workers, and also requests from relatives and neighbours.

Mental distress still carries a stigma of shame, and in many cases the most difficult task is to persuade the patient that he is ill, or to persuade his family that their loved one is not about to be put away somewhere unpleasant. On the other hand, there are some who rather wish their sick relative could be conveniently accommodated elsewhere, without delay, preferably free of charge. The cult of the welfare state has affected the way some clients expect everything to be provided for nothing; if a pension or a grant was awarded they would ask what else might be claimed, on the occasion of my next visit.

Tyrwhitt House, Leatherhead is one of only three short stay homes run by the Ex-Services Mental Welfare Society. I have often related the story of a group of war veterans sitting together in the lounge and grumbling, as only old soldiers, sailors and airmen know how. They complained collectively how they had been ship-wrecked, machine-gunned, taken prisoner, dive-bombed. . . . The majority were Second World War conscripts, so I jokingly patted one on the back and said:

'Well, are you glad you joined?'

'Wouldn't have missed it for the world!' came the reply.

'How is that,' I said 'if you had to put up with all the horrors you have just been complaining about?'

'Ah, well', he replied, 'you see, we tend only to remember the good times. Tragedies such as the death of a comrade, we try to put out of our minds.'

This book tells their story – the horrors and fears they could not leave behind on the battlefield.

I have brought my narrative of Service life up to 2010 with a brief reference to the fighting in Iraq and Afghanistan in the last decade. The latter is still a serious source of casualties while the former is now the subject of an on-going Enquiry in which there is considerable public interest and concern.

Only the other day it was publicly announced that British fatalities in Afghanistan had outnumbered those we sustained in the Falklands.

Note

[1] The term 'ex-Servicemen' includes, of course, ex-Servicewomen wherever appropriate. The worst aspects of modern warfare and its consequences can affect everyone regardless of sex. Civilian casualties included men and women alike, but it has been the man, for the most part, who actually fought the battles of the twentieth century. And predominantly they have been the most seriously affected by the resulting traumas as a result. In the Second World War about 10 per cent of the armed forces were women. After the war, as this book demonstrates, women made a vital contribution towards the rehabilitation of many of my clients.

1

'Shot at Dawn'

In February 1993 the *Daily Telegraph* reported that the Prime Minister had refused to recommend posthumous pardons for those British soldiers shot for cowardice, desertion or other military offences during the First World War. Mr Major commented that we cannot rewrite history by substituting present-day attitudes and values for those of the past. Nevertheless, the Prime Minister recognised that those who had deserted had done so in the most appalling fighting conditions and under terrible pressures.[1]

The *Daily Telegraph* report arose from the case of a Private Harry Farr of the West Yorkshire Regiment who was shot in October 1916. He had only recently returned to the front after five months in hospital recovering from shell shock. His medical condition had been ignored. Relatives were so ashamed of his cowardice that they had not taken the matter up until they were informed that Harry Farr's grave in France did not exist.

I wrote to the Editor commenting on the published photograph, which showed a smart soldier in his blue dress uniform wearing a lance corporal's stripe and a good conduct badge for exemplary Service. Private Farr was a pre-war regular soldier who had seen a number of years Service before being sent to France. He had not shown previous signs of cowardice. After displaying signs of mental illness, which the Army clearly recognised by sending him to convalesce, it was a grave error of judgement to impose the supreme penalty on a man who had become ill for the second time.

The details of Farr's court martial will not be published until after the millennium. When cases such as Farr's were mentioned in Parliament, during and after the war, it was revealed that court martial proceedings could not be disclosed without the consent of the accused. And as the victims were dead anyway, that was the end of the matter.

In my letter to the Editor I noted that had Private Farr been a more recent case of a nervous breakdown whilst in the Forces he would have been considered for a War Disability Pension, medically discharged with a Service pension reflecting his years as a regular soldier, and steps would have been taken towards rehabilitating him after release. Had he died later, his widow might have qualified for a portion of his pension. Indeed, it might have been possible to rehabilitate him elsewhere within the Army to enable Farr to continue to serve his country. At all events, his life would not have been wasted and his family would not have had to bear the burden of his disgrace. Sadly, many of the ex-Service organisations who help such cases were formed only a few years after Farr died.[2]

*   *   *

The loss of the American Colonies in the eighteenth century led to feelings of public contempt for the Army, defeated by an irregular collection of disgruntled settlers, so it seemed. Wellington, The Iron Duke, decided it was best to rule his troops with a rod of iron, and he described them as 'the scum of the earth'. Men who could not be trusted not to desert were kept on continuous overseas Service. In the Peninsular War against Napoleon, harsh flogging was thought to have saved lives by preserving some semblance of discipline. Wellington acted resolutely when plundering had led to drunkenness, followed by desertion. Courts martial meted out swift justice when absentees and stragglers outnumbered the casualties of battle.

After the Crimea, the Army Discipline and Regulation Act of 1879 became the Army Act, which led to the Manual of Military Law. Queen Victoria's army improved in loyalty and discipline as the regimental system developed and the men began to know and respect their officers. By 1914 the British Expeditionary Force (BEF) which went to France was perhaps the best organised, equipped and disciplined regular army in Europe. However, it laboured under a set of rules which clearly owed its origins to the Duke of Wellington himself.

The Manual of Military Law was summarised in a little red book given to all recruits and which listed twenty-five capital offences, including desertion whilst under orders to proceed on active Service. All twenty-five were actually punished in the First World War, and there were seven principal offences listed:

1. Shamefully abandoning or surrendering a garrison, a place or a post, or compelling another to do so.
2. Shamefully casting away arms, ammunition or tools in the presence of the enemy.
3. Treacherously corresponding with the enemy or sending a flag of truce through cowardice.
4. Supplying arms to the enemy or shielding an enemy who was not a prisoner of war.
5. Taking Service with the enemy whilst a prisoner of war.
6. Knowingly, on active Service, committing any act to imperil His Majesty's Forces.
7. Behaving cowardly or inducing others to do so.

Item six included leaving one's post for plundering, breaking into military stores, striking or attacking a sentry, impeding the Military Police, plundering houses or supplies, giving any false alarm, betraying a password, being asleep or drunk on sentry duty or leaving a sentry post. The charge also included taking part in a mutiny on active Service, disobeying an order, striking an officer, desertion or encouraging others to desert, and any other treasonable offence.

The first official death sentence was carried out in the first six weeks of the war when a soldier took off his uniform and was caught wearing 'civvies' in France. By Christmas 1914, 85,000 had been killed or wounded in the BEF – all trained regulars. It was felt that only fear of death would keep their reinforcements at their posts. Nine were sent home with 'shell shock'. This was a new statistic, of great significance. Poison gas and a seventeen-inch naval gun at Ypres faced Kitchener's new army of civilians in uniform. There was also the flame-thrower, and when men retreated from it a company sergeant major drove them back towards the flames with his revolver. Survivors going home on leave to 'Blighty'[3] began to tell stories previously censored in letters home.

By June 1915 MPs began to ask questions and, as a result, the Army Act was amended to allow for some death sentences to be suspended; some units were finding themselves short of men. The amendment also permitted a sentence to be remitted for good conduct or gallantry. However, there was the case of a soldier blown up by a mine being sent back to the trenches after three weeks' respite; a bundle of nerves, he ran away from his trench after six more days of continuous bombardment. He had volunteered for the Middlesex Regiment in 1914 and he was executed in March 1916.

Parliament became concerned about the lack of legal representation for those accused of capital crimes; there were complaints about the sending of open postcards to notify a family of a shooting, and the lack of any system of appeal. Meanwhile, the authorities were in a dilemma – adverse publicity might affect recruiting.

In December 1916 the Royal Army Medical Corps (RAMC) set up a treatment centre for nervous disorders and in the First Battle of the Somme several thousand men were stood down, suffering from nervous anxiety. Some were sent home, but in the battle zone the authorities could not distinguish between psychoneurosis and malingering. The Army simply could not afford wasted manpower. Cases withdrawn from battle might be labelled 'Not yet diagnosed – Nervous' and a number of so-called 'neurologists', who were in fact psychiatrists, were appointed to various medical centres. Some patients arrived labelled 'G.A.K.' – God alone knows!

There had been occasional cases of shell shock on the North West Frontier of India, and in the Boer War some soldiers had been diagnosed as insane; but there had been nothing in recorded history to equal the distress of the trenches in France. It was concluded that shell shock was really another form of civilian neurosis. But in the Army, men faced death and separation from their families. They all felt guilty over the death of their comrades. Some were exhausted from lack of food and rest, some lacked adequate training, but even seasoned regulars broke down after prolonged trench warfare. Junior officers could not cope with having to lead young men to an almost certain death. Some of the most vulnerable occupations included tank crews, machine gunners, mine layers and forward observation post crews. The latter might be floating in an airship over enemy lines for hours on end.

As the war proceeded medical officers aimed to prevent too many men from being evacuated home. Some soldiers returned to duty within 48 hours and even some of the more serious cases could be reclaimed for general duty in general Service within the Army. The term 'shell shock' appeared to apply to psychoses, psychoneuroses, and mental defectiveness alike. The military hero concept could never admit to any psychiatric problems. There was also a puritanical tendency to regard all illness as an excuse for evading one's duty, not to mention the 'stiff upper lip' philosophy.

An acceptable neurosis in wartime was thought to be the result of a physical injury such as a bomb blast; anything else must be cowardice in some shape or form. It could be a matter of chance whether a shell shock victim was regarded as a deserter, a coward, or 'mentally wounded in action'. It was not until June 1917 that casualty clearing centres had an experienced medical officer able to treat and dispose of all nervous or mental cases.

The Manchester Regiment was on the Somme. Morale was high because they were an 'old pals' unit, men from Lancashire who had volunteered together. One of their comrades had been court martialled for desertion when, in the gruelling battle, he had become separated from his regiment, by accident. In fact, he had attached himself to another unit but, when reporting back to his own lines, he did not get a note signed by an officer to explain his absence. The prisoner and guards were cheerful, and shared their food. The prisoner had a previous good record in battle, and had expected little more than a spell on fatigues in the cookhouse, peeling potatoes.

> The battalion was formed up into a hollow square inside a disused quarry. An ambulance drew up on the edge. An old kitchen chair was placed in the centre. The crime and its punishment were read out. The sentence had been confirmed by General Haig, the Commander-in-Chief. The lads could not believe what they were

hearing; after all, they were all old pals and had faced enough hazards from the Germans every time they went over the top. Where was the reason for killing one of your own pals?

An officer, a sergeant, a firing squad of six, and four stretcher-bearers now took over. One rifleman had a blank round so that no one would ever know who actually killed the condemned man. The prisoner was tied to the old chair with a white cloth over his chest for the firing party to aim at. The officer dropped his handkerchief as a silent signal to the men . . .

The officer then ran forward to the chair and made sure the man was dead with a shot to the head from his revolver. The victim's hair was standing on end from sheer terror when he died. The firing party and bearers dug the grave and buried the body, wrapped in an Army cape. A wooden cross was inscribed with his regimental number, rank, name and the word 'killed' – not 'killed in action'. No one in the regiment could see what good this had done and the men were gloomy for days afterwards, deep in thought and apprehension.

This eye-witness account of a shooting was described to William Moore (*The Thin Yellow Line*, ch. 1 [Leo Cooper, 1974]).

By 1917 some men had become fatalistic in their depression. They no longer sang. High casualties amongst the officers meant that they did not always know their own platoon commander. Families were reluctant to complain when their sons were executed because of the shame of it all. There were, however, a few glimmers of hope. In time, a few officers with some legal experience found themselves on courts martial where they began to ask awkward questions. Other questions were raised in Parliament when doubts were expressed as to whether a defending officer had been present at some of the hearings. Another aspect was that the relatives of those shot after the 1916 Irish Rebellion might ask difficult questions if military trial proceedings were published for everyone to read.

Towards the end of the war Haig wrote to the Under Secretary of State for War that when there were doubts about neurasthenia or shell shock, a condemned man should appear before a board which included an officer with neurological experience. But this did not bring back those who had already been shot. In 1918, 38 men were executed for surrendering to the enemy. Young boys, many of them under age, found long marches with heavy packs too much for them; they suffered from neuroses and some were found asleep on sentry duty.

The Americans found that heavy First World War artillery fire could lead to shell shock, and nervous cases in military hospitals would hide during an air raid or even if a door was slammed. By contrast, men with physical wounds alone would joke about them. Although there had been a tradition of shooting deserters in the American Civil War, in the First World War suspected malingerers were seen by a consultant psychiatrist, and the Commander-in-Chief had to confirm sentences passed for mutiny or desertion. The President could commute a death sentence for other offences. The United States admitted that, before neurosis had been confirmed among some soldiers, a few had been executed as malingerers. Some had committed suicide following conviction.

In the British Army, as the war proceeded, some cases which would have led to an earlier shooting were now considered more carefully; few actual malingerers were found. The primary aim of the authorities was to prevent too many cases of evacuation home. Some soldiers were sent back to duty within a few days. Those with longer lasting phobias were a greater problem, but even they could be reclaimed for general Service duties, although those below normal physical or mental ability could not. The long Service regular who lived in fear of blighting his chosen career needed special consideration. Those temporarily dazed and accused of desertion might be observed for at least a month and there could be a fine line of judgement between the coward and the genuine casualty of the most horrendous war in history. Those attempting to treat or judge these unfortunates had seen nothing like this scale of misery on the battlefield.

The general consensus was that all cases of shock had to be caused by an actual physical shock and this theory left out cases of men living on their nerves who could barely cope with Army life without the double fear of death from behind and in front. Military policemen and men in firing squads were deeply affected by what they saw. Those who were shot were considered scapegoats. (No one ever wanted to be a member of a firing party. However, a decision to classify executed soldiers as 'died on active Service' helped avoid some of the outcry from the public.)

Comrades were shocked and saddened, but there was a remarkable acceptance of it all as one of the tragedies of war. There is no evidence, and there was no evidence at the time, that shootings were essential to maintain a high level of discipline in combat. Far more important was regimental spirit, and good leadership from junior and middle-ranking officers.

The highest percentage figure for desertion was 20 per cent for 1915. This was the year which included Ypres and Gallipoli, and when Kitchener's volunteer army had lowered the standards of acceptability for military Service. Thus, the death penalty had little effect here. Better selection paid off and the rate of desertion was down to 7½ per cent by 1918.

All humans are faced with constantly changing conditions in life to which the mind adjusts by changing its senses of perception, co-ordination and physical reactions. Abnormal circumstances may not give the body time to adjust and a situation to which the body cannot adjust is said to be *traumatic*. There are times when the body cannot cope because of extreme fatigue and it is simply unable to react normally at all. Thus there must be reasonable doubt as to the mental stability of the 11 per cent of the condemned men who were executed during the First World War.

*   *   *

I began working with mentally disabled ex-Servicemen in 1984 and two cases of Great War veterans come to mind. Numbers are decreasing all the time and statistics for all disabled personnel in receipt of a War Disability Pension, in 1993, showed nine cases of men aged over one hundred, and 687 then in their nineties.

Fred lived in a semi-basement in a fashionable seaside resort in the West of England. Only the very tops of his windows showed above the ground, where crowds of holidaymakers went to and fro. It was damp inside and permanently gloomy save for the light which came from a single bare bulb. Fred was born in 1894 and I saw him in 1986 and 1987. He had served in the Battle of the Somme but eventually cracked up and he relied on a small war pension for the rest of his life, only able to do labouring jobs. He said he never married and had no known relatives. I saw him outlive his next door neighbour, a lady in her eighties, with whom he had eaten his meals-on-wheels for many years. The flat was none too clean and Fred had no surviving friends who bothered to visit him.

Fred would have been suitable for a place at Kingswood Grange, the Society's permanent home for old soldiers. However, they had tried him in a local authority home where, typically, there were many elderly ladies. He did not relate well to the others and so he had returned to the flat. He avoided talking about the possibility of going to Kingswood Grange; he related how he had been burgled by a teenager who had stolen his electricity money and Fred was awaiting the outcome of the court case. The money was never recovered because the youth, who later pleaded guilty, was unemployed. This angered Fred who remembered that there was supposed to be a land fit for heroes after *his* war!

I suggested to his GP that we should try to get Fred into a home for elderly men, such as ours, but we never succeeded. He was very talkative and clearly needed company. There was even some talk of the Council pulling the building down, where he lived, in order to widen the road; but Fred would not discuss this either. It was not possible for me to call very often, but Fred badly needed a friend he could trust. Sadly, he valued his independence so much that he was reluctant to ask anyone for help. He died shortly after being transferred to a nursing home in 1989, aged 95.

My second example is the case of **George**, who was born in 1896 and lived on the Borders of Wales. I saw him first in January 1986. He joined the Royal Garrison Artillery in 1915 and was discharged in June 1918. George was wounded on the Somme with shrapnel in the forehead and the left side of his face. He was medically discharged, under King's Regulations, and awarded a small disability pension. He worked on building sites between the wars, married, and later worked in a munitions factory during the Second World War. After the war he was a railway line maintenance man.

George deteriorated mentally after his wife died in 1980. When I saw him he was in good physical shape for his age, living in a single person's Council flat which was remarkably clean, and where he did a little gardening for himself. Relatives did not bother to keep in touch, as they regarded him as eccentric. George was not short of money for his immediate needs. He was very fortunate in that he had been able to work for most of his life.

I took George to Tyrwhitt House, the Society's home, for a short break in February 1987 and he thoroughly enjoyed the company, fitting in well. However he was keen to get back to his own small place. A year later he was encouraged by his social worker, with the best of intentions, to go into an old people's home, and died there a few months later. It is often the case that elderly patients do not survive being uprooted from their familiar surroundings. Where possible, it is best to enable clients to remain in their own homes.

The Great War is now 80 years past, and clients, in their nineties, were typical of an age when personal independence was of supreme importance. They deserve more help, of the right kind, than they ever received in the past.

Postscript: It was recently revealed in a BBC programme on Radio 4 that Harry Farr's widow was visited by the vicar of the church the family attended. He had received a letter from the Chaplain of Harry's Regiment who had been present at the execution. As he faced the firing squad, Farr insisted on not being blindfolded and, as he died, he looked the riflemen squarely in the eye. The Chaplain concluded, 'This man was no coward'.

Notes

[1] Statistics of the Military Effort of the British Empire During the Great War – Part XXIII 'Discipline' reveal that 346 members of the Forces were shot: 322 had been in France and 263 had been for desertion. Other shootings were for murder, cowardice, quitting one's post, striking a superior, destroying an order, mutiny, sleeping on duty and casting away arms. Trials for cowardice increased as the size of the Army increased:
126 in 1915, 136 in 1916, 161 in 1917
There were 3894 cases of self-inflicted wounds.
[2] The Soldiers' and Sailors' Families Association was founded in 1885. The Ex-Services (Mental) Welfare Society dates from 1919, the 'Not Forgotten' Association from 1920, and the (Royal) British Legion from 1921.
[3] 'Blighty'. Soldier's name for England or the homeland widely current in World War I, but was well known to soldiers who had served in India long before. It is the Hind. *bilayati*, foreign, from Arab. *wilayet* meaning

'provincial', 'removed at some distance'; hence adopted by the military for England. Its widespread use during the Great War is illustrated by three popular songs of the time – 'There's a ship that's bound for Blighty', 'We wish we were in Blighty', and 'Take me back to dear old Blighty, put me on the train for London town' (*Brewer's Dictionary of Phrase and Fable*).

2

The Early Activities of Ex-Service
Charitable Organisations

In order to follow the development of the ex-Service charitable organisations, we must first look backwards in time, and focus not only upon the men, but also on their wives and children. In the past, soldiers in barracks and lodgings at home had wives and children, although the authorities did not always recognise the standing of a soldier's immediate family.

By 1885, some wives were on the strength of the Regiment, home and overseas, on peace-time garrison duties. There were also other wives who married, without permission; and, officially, these 'wives' did not exist at all. Generally, sergeants' and warrant officers' wives were officially recognised. The rest were not. Even those on strength could be turned out of barracks if the unit went on active Service. If the husband died, other than on active Service, there was no pension. The unofficial wives got nothing at all, except a stoppage of threepence a day from their husbands' pay which was paid to them. The Admiralty did not recognise that sailors had any right to marry at all.

As the second expeditionary force set sail for Egypt in February 1885, a major in the Royal Warwickshire Regiment, later Colonel Sir James Gildea, wrote to *The Times* appealing for funds and for volunteers to help those unfortunate families left behind. The climate of opinion at home was not welcoming and nineteenth-century England had been civilian in attitude ever since Waterloo. Soldiers were written off as a disgraceful rabble while sea power to defend the Empire was all important. The patriotism which arose during the Crimean War became tempered by the fact that the Army was found to be inefficient. The Queen herself, with her

German connections, was largely unaware of the potential dangers to Britain from Europe. International politics were of no great interest to the Victorians. They had their Navy and the Empire was run as a highly successful world trading enterprise. State intervention at home was something to be avoided and, all things considered, the Army was not needed.

In spite of all this, shortly after the first *Times* letter, a committee of ten ladies had been formed, funds were raised, and a small allowance became available for soldiers' wives and elderly parents. Help was provided towards families finding work in order to help themselves. Local committees were formed in garrison towns in order to help each other, and the Princess of Wales (later Queen Alexandra) became the first President of the Soldiers' and Sailors' Families Association.[1] Clothing branches, county committees, overseas branches, and their renowned home nursing Service all followed quickly. SSFA nurses were to pioneer

H.R.H. The Princess of Wales: 'Do not be uneasy about your wife's comfort, Tommy; I will look after her and the children,' from *Punch* during the South African war, reproduced by kind permission of SSAFA Forces Help.

the very idea of a health visiting Service as we know it today.

The success of SSFA was perhaps the greatest incentive towards raising the necessary funds during those early years. All of this increased public awareness of the need for charitable enterprise to support the Services and was to anticipate the benevolent work with the men themselves, after discharge, which arose, on a large scale, after the First World War.

There were other developments, nationally, as the Victorian age gave way to the Edwardian. Victorian women with time to spare had run local charities for the poor, providing shoes, coal and blankets; and the early feminist movements of those times gave some point to their activities. There was certainly a need for charitable work with the poor, when it was found that 60 per cent of recruits for the Boer War were unfit for Service. Such revelations were eventually to stimulate governments in the next decade to initiate school meals, schools medical examinations, the Old Age Pension, employment exchanges and a basic National Insurance Scheme for sickness and unemployment benefits. Charities were also active in the arts and public libraries, a role later taken on by the local authorities.

Standards of living among the working classes improved before, during, and after the Great War, in spite of recession in the 1920s and 1930s, when there was always a million unemployed. The fact that real wages were 30 per cent higher in 1938 compared with 1913 meant that, although there was a tremendous gap between the very rich and the very poor, the skilled workers could afford some luxuries and support charities. During this period, from around 1890 to the Second World War, much voluntary welfare work was done by the Salvation Army, and the YMCA. And there was also a National Council for Social Services which organised activities for the unemployed.

Of importance was the increasing contribution of women to public life and charitable works. Lady Astor took her seat in the Commons in 1919 following a world war in which women had played an increasing part in the community, while the men were away.

On 1 November 1918 the first meeting of a charitable society to establish 'The Fellowship of Reconstruction and Welfare Bureau' was held. There were four ladies present, and three ladies and two gentlemen sent their apologies. This shows that the women by far outnumbered the men, also that some of them perhaps did not, then, attach to this venture the importance it deserved. From this beginning, the Ex-Services Welfare Society was established in 1919. Its name was not changed to The Ex-Services Mental Welfare Society until 1958.

The Society's functions developed with a three-fold aim of the relief of distress, the provision of employment, and professional medical assistance. There was a head office in London to which scores of veterans resorted for gifts of clothing, small grants of money, and medical advice. A recuperative home usually existed in the London area or in East Anglia between the wars. There was also much activity in Leatherhead where, under the Presidency of Sir Frederick Milner, a large house, known as Milner House, was purchased for £5,500 in 1926.

The following year, an industrial colony was set up at Milner House, mainly to produce 'Thermega' electric blankets. There was

Milner House:
Picking tomatoes
for market.

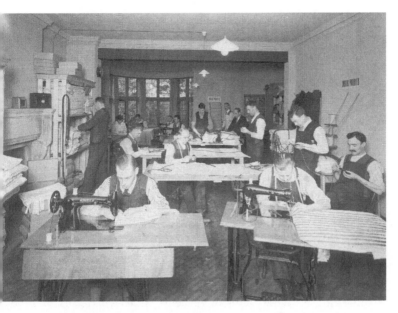

Milner House: Manufacture of "Thermega" electric blankets.

also a successful market garden. This was a major step forward and the house provided permanent accommodation with the factory literally at the bottom of the garden. No treatment was available on site, but there was work available for many of those who were able. The only problem was that mentally disabled workers are not always one hundred per cent reliable. They have good and bad days, which may be brought about by swings of mood. Any commercial organisation employing these veterans may well encounter production problems, even though their work is satisfactory for most of the time, when they are feeling well.

Twelve cottages were built at Milner in 1930 for occupation by married couples, where the man worked at 'Thermega'. Many of the cottages were provided by other ex-Services charities, and thus the idea was born of a small industrial, therapeutic community. A social club was set up at Milner House itself.

The success of this venture was marked by the arrival of the Duke and Duchess of York in 1932. The Duchess, now Her Majesty

Milner House: Men working in the fields, showing cottages built for families in the 1930s.

Queen Elizabeth the Queen Mother, is still very interested in the work of the Society. She became Patron in 1942 and has remained so ever since. Milner House was frequently visited by other dignitaries, crowned heads and politicians, and acquired an international reputation. Admiral Sir Reginald Tyrwhitt became President in 1933, and a treatment centre was set up at Beckenham. Tyrwhitt House, to which he gave his name, was purchased in 1945. By 1939, the small Ex-Services Welfare Society, dealing with the mental casualties of war, was active on all fronts. Public awareness of this work was slight, and I suspect that the vast majority of the mentally disabled from the Great War were largely unaware of the Society, or were reluctant to discuss their problems with anyone, least of all their families or relatives. Public attitudes towards mental illness were generally intolerant and a sense of shame might descend on the family of anyone unfortunate enough to require residence in an asylum.

Similar attitudes might well have been the case when a shell-shock victim who could not "pull himself together" found himself

Milner House: The visit of the Duke and Duchess of York in 1932.

being interviewed by a psychiatrist at Head Office in Craven Road. He might be offered clothing and a few shillings in his pocket, and then a short stay, perhaps a couple of months, at Eden Lodge Beckenham, or a more permanent stay at Leatherhead, with job prospects.

The 'Not Forgotten' Association was founded on 12 August 1920 by Miss Marta Cunningham, an idealist with a tremendous personality and determination. It all began with a visit to a hospital near her home where she found 600 wounded soldiers, silent white-faced men. She realised there must be many thousands of these shattered youths all over the country, with only the four walls of their houses of pain to greet them, day after day – paying in blood and prolonged agony the price of the world's and our safety. The thought was unbearable. She sped home and within half an hour, six astonished hostesses were ready to open their homes and hearts to any patient able to leave his bed. Miss Cunningham enlisted the aid of three wonderful presidents – Countess Beatty, Countess Haig and Lady Trenchard – representing the Royal Navy, the Army, and the Royal Air Force.

She then proclaimed to the press the name of The 'Not Forgotten' Association.

In the first year ten thousand patients were entertained in private homes, and soon afterwards the enterprise was supported by the Royal Family. A summer garden party for disabled and wounded ex-Servicemen has been held at Buckingham Palace since 1921, and there have also been Christmas parties. Top artists from the entertainment world have given freely of their time to welcome the guests. Hospital patients are taken on outings, there is a holiday scheme, and television sets are provided for any lonely ex-Servicemen who qualify. I have recommended dozens of clients for help from the Association and I cannot recall anyone being refused. This is the only organisation, to my knowledge, about whom this can be said.

There is a guest house in Bournemouth, run by an ex-Service family with a disabled relative from the Royal Marines, where my clients and their wives have received tremendous support; and these holidays are still being paid for by the Not Forgotten Association. Another client of mine who cannot read or write, has recently acquired a television set, with installation, licence and rental all paid for.

The aim of the Association was defined as 'To supplement for the seriously war wounded and disabled the work of public authority and organised charity by making the lives of hospital inmates a little lighter, brighter and easier. To bring them entertainment, amusement, social intercourse and certain minor luxuries and creature comforts. To break the march of the slow, long days and allay that yearning for a new face which knocks at the heart when you have the same faces always with you. 'The "Not Forgotten" set out to take these men and women out of themselves, to make them feel that they mattered.' This is still the aim today. I have been able to recommend dozens of clients for help from the Association and I cannot recall anyone being refused.

The 'Not Forgotten' Association was established a year before the British Legion (now Royal British Legion), which grew out of the National Federation of Discharged Sailors and Soldiers, the National Association of Discharged Sailors and Soldiers, and the Comrades of the Great War. The Legion was to become the

largest British ex-Service organisation providing welfare
Services, employment, social facilities, and practical help towards
obtaining War Disability Pensions.

In addition, Regimental associations increased in size and
number after the war although the Royal Artillery Charitable
Association was founded as long ago as 1830. The Regular Forces
Employment Association, established in 1885, had its origins in
the Army and Navy Pensioners' Society founded in 1855, and the
Corps of Commissionaires (1859).

Before the First World War, War Pensions were vested in the
War Office for army officers, the Chelsea Commissioners for
the Army for other ranks, and the Admiralty for naval officers
and seamen. Pensions for wounds and injuries were regarded as
a discretionary grant. The situation changed dramatically when
the size of the Army trebled, consisting mainly of temporary
soldiers or civilians who would, it was hoped, soon return home
after hostilities. What was needed was a type of industrial
compensation scheme, legally claimable, echoing the rudi-
mentary National Insurance Scheme set up by the Government in
1909. Consequently, a Royal Warrant of 1915 provided for a
pension of 25 shillings a week for total disablement, with
allowances for partial disablement; in addition there were children's allowances, parents' and widows' pensions.

A Royal Warrant of 1916 recognised the principle of an aggravated injury – one which was made worse as a result of Service in
the Forces. By February 1917, with half a million casualties to deal
with, the Ministry of Pensions was established; the present system
of War Pensions Committees was set up in 1921. The Royal
Warrant of 1917 changed the basis of compensation from loss of
earnings to the degree of disablement suffered.

Thus there was a network of assistance for the ex-Serviceman
fairly soon after the First World War and this state of affairs has
remained broadly the same ever since. Money has come from
public subscription from a grateful nation, often for some specific
cause (a branch of the Services, the Royal Air Force, for example),
or for a special occasion. Governments have provided pensions
and the Servicemen themselves have contributed a day's pay per
annum to their own regimental charitable funds. Because it has
been a voluntary system, largely with many funds operating in

isolation, its effectiveness must inevitably remain patchy. Few people are fully aware of the totality of what is on offer.

There have always been the problems associated with making one's case known, with the usual inhibitions and stigmas attached to accepting charitable aid or being labelled as a cripple. These attitudes have particularly affected working with the mentally disabled and more will be said about this later on, in chapter 7.

Between the wars, the voluntary charities made a point of issuing clothing to needy ex-Servicemen, often with few questions asked. There were always some who would take their new things and sell them, spending the money in the nearest hostelry. Life could be dull, so this was understandable.

In the years after the War the debate on shell shock and the retention of capital punishment continued; all political parties had members who supported it. Flogging in the Forces was abolished in 1881 and, perhaps surprisingly, it was not found necessary to bring it back during the South African War. Field Punishment Number One was carried out when military criminals were tied to a fixed object for public humiliation; it was widely used in cases of drunkenness. It was abolished in 1923 and had been the military equivalent of putting prisoners in the stocks during the Middle Ages.

In 1922 the War Office set up a Committee of Enquiry into Shell Shock under the chairmanship of Lord Southborough. It was to consider its origins, treatment, and avoidance through training and education. Capital punishment for 'cowardice' was not to be ended immediately.

The Royal Army Medical Corps, soon after its formation, established a specialist subject, 'psychological medicine' in the Captain to Major Promotion Examinations of 1898. RAMC opinion was perhaps crystallised in 1920 when Dr Stanford Reed of the Royal Victoria Hospital, Netley, wrote that military misfits should be examined together with frequent offenders and court martial defendants.

The Southborough Enquiry was very revealing. A Ministry of Pensions neurologist (we would describe him as a psychiatrist today) refused to accept a man for a War Pension on the grounds that he had simply broken down mentally whilst on active Service. He said he would accept a case where a soldier had been physically injured, blown up or buried. It was reported that the greatest cause of shell shock was loss of sleep. Another doctor described it as 'prolonged cowardice'. Shell shock was said to be a misnomer; it was not a new idea, but the stresses of war led to cases being exaggerated in both numbers and severity. There were said to be three types of shell shock caused by, first, concussion, or the effect of a heavy blow to the head; second, the emotional shock of a terrifying event which might be triggered off by a more trivial event; and third, mental exhaustion from a combination of great nervous strain and prolonged physical hardship.

Cowardice was recognised to be beyond an individual's control and that a consultant should decide where this was in doubt. Special consideration should be given to men of proven bravery who later collapsed. There was some justifiable criticism of medical examinations for their lack of any psychoanalytic component. It was strongly emphasised that no one ought to consider shell shock as an honourable means of escape from battle. Slight cases of neurosis should be rested near an active Service area before any return to the fighting. The Committee recommended the use of a consultant psychiatrist at courts martial for cowardice to advise in distinguishing it from genuine psychoneuroses.

It was argued that the Army trains men not to be afraid and then the country imposes the death penalty to give them something to be afraid of – an illogical state of affairs. However, the British public were not very interested in the arguments for or against the death penalty. The War was over and that was that.

A. P. Herbert's play 'The Secret Battle' was based on the death of Sub-Lieutenant Edwin Dyett, a naval officer in France when sailors were being used as infantry because of the desperate shortage of fighting men. Dyett had told his commanding officer that he was under strain and could he please go back to sea. The CO persuaded Dyett to withdraw his application for a transfer, and was soon afterwards himself killed in action. Dyett was found behind the lines and was executed for cowardice at Christmas 1916. Herbert's character, Harry Penrose, was described as 'one of the bravest men I ever knew'. Herbert believed in the death penalty but that Penrose was not a coward and his trial was unfair.

In fact Dyett met his defending officer for the first time only four hours before the hearing took place.

In 1928 the death penalty was removed for sleeping on duty, striking an officer, plundering and destroying an order. The death penalty remained for cowardice and treason. Duff Cooper, Financial Secretary to the Treasury, stated, 'There is no such thing as cowardice, and it ought to be punished severely'! The Southborough Committee in 1922 had reported that good morale is the most important object of military training and that by such means the soldier would be protected against the occurrence of shell shock. It was felt that all soldiers had an endurance threshold of 250 days in action, after which they would crack up. Service chiefs have, by tradition, wrongly supposed that every decent soldier is capable of resisting physical and mental stress – a fallacy that has lost battles and cost countless lives, over the years.

In 1930, in spite of opposition from Viscount Allenby in the Lords, the Commons agreed that a soldier who broke down under the strain of war would no longer be shot. Death was only retained for treason and espionage.

In 1939 there were still 120,000 veterans in receipt of a War Pension for psychiatric invalidity.[2] In July of that year the Ministry of Pensions convened a conference to consider cases of neurotic breakdown in time of war. There was a fear of an epidemic of 'war neuroses' breaking out should hostilities commence, as indeed they did. It was recommended that no one should be discharged for neurosis, and no one receive a pension for war neurosis except in special circumstances. The aim was that there should be set up rehabilitation centres for cases of neurosis, to enable soldiers to return to active duty as quickly as possible. Before 1939, all cases of psychosis, or serious nervous disorder, were discharged. The only cases not discharged were temporary traumatic or toxic disorders, or mild psychoneuroses.

At the outbreak of the Second World War the ex-Service charities were still caring for the worst cases from the 1914–18 war and, as the country prospered, money was donated from all sections of society. The death penalty had gone, for alleged acts of cowardice, but many of its advocates were still in positions of influence in politics and in the Services. History would repeat itself as a small army of regulars was supplemented by millions of civilians in uniform. Suffering was to become widespread among the civilian population, while active Service conditions for the Navy and the Air Force were to be quite unprecedented in scale and in the degree of horror which had to be faced.

Notes

[1] The Soldiers' and Sailors' Families Association (later Soldiers' Sailors' and Airmen's Families Association in 1919) was to become the first large Service charity, which ultimately grew into an organisation operating on a national scale. Some of its work was to spill over to the benefit of the husbands themselves, both serving and ex-Servicemen.

[2] Defined as neurasthenia, shell shock, effort syndrome, epilepsy and insanity. R. H. Ahrenfeldt, *Psychiatry in the British Army in the Second World War* (Routledge and Kegan Paul, 1958), p. 10.

3

The Second World War: Casualties of Dunkirk, North Africa and Italy

Most members of my generation remember perfectly well what they were doing on 3 September 1939. I was a schoolboy, a member of the Boy Scouts and on the first Sunday in the month, excepting August, we were expected to attend church in our uniforms, sitting on the front row through a Service not really intended for us. Hitler had invaded Poland on Friday the first, at which point my father began to make blackout shutters out of plywood which fitted inside the windows after dark and might protect us against flying glass. Mother had already bought some black Italian material with which to line the curtains.

Church started at ten-thirty and the Minister said, "I suppose you all will wish to go home to hear the Prime Minister on the radio at eleven so there will only be a very short Service." We said The Lord's Prayer, we sang Onward Christian Soldiers followed by God Save the King, and that was that. As boys, we were pleased to have been spared sitting through the inevitable sermon about Sin; after all, far greater sins were being committed in Europe, we thought. I arrived home at ten to eleven. Father was still sawing up pieces of wood and mother had uncovered her treadle-sewing machine, ready for action. The radiogram was switched on, Neville Chamberlain made his historic announcement on the hour and, to my great surprise, it made no difference to us! There were no German bombers overhead, we had our Sunday lunch and, apart from putting up the blackout, it was just like any other Sunday at home.

We were to be only just too young to actually fight in the war although my generation were to take over in the Forces from the men who had actually taken part, at the very end of hostilities. Many of us occupied Germany, we retook Singapore from the Japanese; some were to lose their lives in Palestine and, later, in Korea. When I was old enough, I joined the ARP, later called the Civil Defence, as an Air Raid Cyclist Messenger. They gave us a blue battledress uniform, trained us in first aid, fire-fighting and anti-gas precautions, besides giving us about twopence a week to pay for our cycle headlamp batteries! I was based at our local church hall which was ready to be turned into a Civil Defence Rest Centre to house and feed the homeless after any serious air raid disaster. Thankfully, the bombs never actually fell on us although there were plenty of scares and near misses as I cycled through the blackout with the German bombers droning overhead. Later on, I joined the Home Guard and eventually qualified for the award of the Defence Medal.

I have made a point of explaining some of this because some two-thirds of the disabled ex-Servicemen I helped were from the Second World War and, because I remembered the events so clearly at the time, I could relate to these men and women in a close personal way. That war was to be fought on such a vast scale that no one could have possibly experienced every aspect of it. Yet, in common with most of my veterans, I do remember something of life in the 1930s and 1940s, together with the joys, sorrows and disappointments of the years which followed the end of hostilities.

Life in the Forces is a world of its own, which those who have never experienced it can only imagine with some difficulty. I survived in the Regular Army for the best part of twenty years while most of my relatives thought I was mad, foolish or just unsettled. When I came home on leave only three questions were asked, 'When do you have to go back?', 'How much longer are you staying in for?' and 'When are you going to settle down?'. The fact that I might have found the Army an interesting, challenging and rewarding way of life never occurred to them at all. The fact that I did not come from a family with a military background meant that I could look at Service life objectively.

In the summer of 1940, a number of Corporation double-decker buses appeared in our street. This was, in itself, an object of some curiosity because buses never ran down our quiet suburban road. Leaning out of the windows were some of the soldiers who had escaped from Dunkirk. They were in various forms of dress. Some had greatcoats on, some had not, some had their steel helmets while a few still carried their rifles. All looked tired and worn out, and these were the ones who had not been seriously injured. A sergeant and a billetting officer knocked on doors asking our neighbours to provide shelter, for an unspecified period, to any soldier who could be given a bed. We had no bed to spare and could not take anyone, but it was at this point that I felt that the war had really begun for me. From then on, we had soldiers stationed all around us; they even occupied a large hall adjoining my school, for the duration. Then there were anti-aircraft batteries on the outskirts of the town.

The retreat from France and the beaches of Dunkirk provided the background for the first significant numbers of casualties with whom I was later to become familiar.

*   *   *

The conference called by the Ministry of Pensions in 1939 had called for treatment, rather than pensions, to be awarded for neuroses amongst Servicemen on active Service. They were clearly expecting an 'epidemic' of such cases. Emergency Medical Service (EMS) hospitals were set up with psychiatric centres to receive any patients regarded as mentally unfit for military duties, as was the case in the Gulf War, 1990–91. In April 1940 every Army Command was given a Command Psychiatrist; military psychiatric out-patient centres were established to act as filters for the EMS system. One advantage of the military psychiatric hospital was that men could be held there without the need for certification, another was that a close watch could be kept over cases where a War Pension claim had been applied for. 'D' Block at the Royal Victoria Military Hospital, Netley, Southampton, the Army's psychiatric centre, was quickly found to be inadequate for wartime requirements. Banstead Psychiatric Hospital in Surrey opened a special wing as a military mental hospital in March 1940 where they also took RAF personnel. At Northfield, Birmingham, in April 1942, the largest special Army hospital opened with 200 treatment beds and 600 places in a training wing for the rehabilitation of neurosis cases, before returning the men to some form of military duty. May 1941 saw a scheme at the EMS Mill Hill where patients were discharged to military units under an 'annexure' system. More will be said about the military hospitals in chapter 5. Psychiatrists went to overseas commands in 1942–3. And in 1942 they were involved in the selection of candidates for commissions.

Ten per cent of all known casualties at Dunkirk were for psychiatric causes, defined as exhaustion, mixed anxiety, or hysteria from situations where men were obliged to struggle to the coast; many lost touch with their units, there was physical hardship and, all the time, there was the risk of being suddenly killed by enemy fire. There were unpleasant sights, German attacks on French refugee columns and, lurking in the background, the possibility of being taken prisoner. For some, the strain proved too much. Hungry, thirsty and desperately tired, tortured by the screeching of dive bombers, exploding bombs, machine-guns and the screaming of the men – by the time they reached the beaches, many soldiers broke down.

It is not the purpose of this book to cover the military and political events of the war in any great detail, nor will I dwell unduly on the horrific aspects of the bombing and fighting, save to illustrate the stories as related by clients. It is ironic that many major works of political and military history have covered this period, while many of my clients – those individuals who were at the spearhead of this history – were, on the whole, reluctant to dwell on those events which had caused their distress. Rather, this book will concentrate on the therapy and other means of combatting the stresses of modern warfare. It will detail the measures that can be taken to obtain compensation and treatment. And it will promote the view that society has a duty to alleviate distress from war-related illnesses wherever they are found.

*   *   *

Stan, who lives on the South Coast, is in his late seventies and is now alone following the recent death of his wife. He refuses a place in a local old people's home because he feels he could not settle in a place predominantly occupied by elderly ladies. This case is typical of many I have found and must surely represent a national problem. Stan's grown-up children do not live nearby, but they are kind and helpful on the few occasions they visit. Stan is a proud man, managing alone with help from Meals on Wheels, the Home Help, and some very good neighbours. He is only partially sighted and relies on a nurse whenever he takes a bath, otherwise he fends for himself. Stan has served his country well and has not been a burden on the State but he admits that he is a poor mixer.

At the outbreak of war Stan was in the Territorial Army, having volunteered at the time of Munich in 1938; his army Service ended in the Military Psychiatric Hospital at Netley in 1942. On the way to Dunkirk, the lorry he was driving was blown up; he then found an abandoned French tank which he managed to drive almost as far as the beach until it too took a direct hit and caught fire. Other soldiers whom he had driven thus far managed to prise him out of the tank. Stan swam out to a rowing boat which capsized under the weight of men trying to escape on it but was able to reach 'Mona's Queen', one of the Isle of Man steamers which hit a mine and sank soon afterwards. Stan eventually reached Dover on an overloaded destroyer. He was suffering from head injuries and it was eventually decided, at Netley, that he was not going to be fit for any other form of military Service.

Stan qualified for a small disability pension, but when he got a job as a window cleaner and failed to correspond periodically with the Ministry of Pensions, the payments were stopped. The window cleaning came to an end after he blacked out several times, and fell off his ladder. Undaunted, Stan became a 'packman', a credit tailor going round the houses selling clothing on 'tick' and collecting the money by instalments. By now, Stan's wife had to help on those days when he simply could not face his customers.

It was the onset of emphysema which finally ended Stan's working days, at which time he was in his fifties. Stan and his wife managed on Invalidity Benefit until he qualified for an Old Age Pension. When Stan's wife died, an application was made, through the Royal British Legion (RBL), for the reinstatement of his War Pension. When I saw him, he was in receipt of an award for an 80 per cent disability. I should add that there is never any back pay beyond the date of application, unless it can be proved that the reason for the rejection or cancellation of a pension was a result of maladministration on the part of the authorities, or that the applicant was mentally or physically prevented from claiming sooner. It is to be hoped that Stan will end his days in an ex-Servicemen's home but, because of his independent spirit, the signs are not encouraging.

There were many others for whom the war finished at Dunkirk. **Leslie** joined up in 1938 at the age of 16 (he said he was 18). He was bombed whilst his unit was retreating towards the coast, and he was medically evacuated, probably amongst the last to have got away from France in this way. Leslie does not remember much about what happened to him but he was discharged from an Emergency Military Hospital in 1942.

He stayed at home with his parents on the outskirts of London for the next three years, afraid to go out. He was especially fearful of the bombing. During this time he received a small War Pension. After the war he did labouring jobs, got married and had children. But by the time he was in his forties, his marriage had broken up and Leslie could no longer work. Whether the one happened because of the other, we do not know but I suspect it was the recurrence of nervous anxiety which ended both. I found Leslie in warden-assisted flats where he had been for decades. He lived by himself, managing to cook for himself, but his body shook from time to time and he could only go small distances in his wheel-chair. His ex-wife and children now live on the other side of the world. Leslie could well qualify for Kingswood Grange.

Bert was a pre-war Territorial who was fished out of the water off Dunkirk, half drowned. He was medically discharged before the end of 1940 and awarded a 60 per cent War Pension for physical disabilities. Twenty years later, it was changed to 'anxiety neurosis' and reduced to only 20 per cent. It was apparently assumed that his physical injuries were better; he had been managing to work as a clerk until soon after his fiftieth birthday when he collapsed at the office.

Bert suffers from a state of constant anxiety. He worries about everything and anything. He reads but cannot remember what he is reading. He worries about going out and getting lost. He loses his balance, loses his breath and suffers from pains in his head. Fortunately he has a caring wife and money is no problem for their immediate needs. In order to give Bert's wife a much-needed break it was arranged for Bert to stay at Tyrwhitt House, but he did not settle. He did not like to be away from his home surroundings. Bert takes a large cocktail of medicines but the local medical authorities seem unable to achieve any improvement in him. I found Bert's case very frustrating. I felt that a course of therapy from a military psychiatrist might get to the root of Bert's problems. I was convinced that Bert's physical difficulties had psychological origins.

I met **Bill** in the Channel Islands where he was attending a day centre for the disabled. He was looked after by his sister and brother-in-law, a situation which had gone on for twenty years, and he was happy there. It had always been maintained that he had received a serious head injury when he was crushed between two tanks at Dunkirk. Bill was a pre-war Corporal in the Royal Tank Regiment and, after returning to this country, he was discharged from Banstead Military Hospital in 1942. He was given a lump sum in compensation for his injuries, and the assumption was that as his injuries only caused a low percentage of disability, it would lead the patient to making a complete recovery and to be able to live a near normal life.

Bill joined his family who were evacuated to Lancashire. He never worked, and they all returned to the Channel Islands after the war. He suffered from epilepsy and blackouts so his mother cared for him until she died. When I saw him in 1986 we were in the process of appealing to have his War Pension reconsidered, and the local War Pensions Officer was instrumental in this. Bill was seen in hospital at Southampton where evidence was found of scar tissue surrounding his brain, which was attributable to his war injury. He had never qualified for anything more than a basic grant from his local Parish because he had never worked. The local Royal British Legion gave him £5 a week. Much of his time was spent at his sister's, in his own room.

Some years later, the War Pensions appeal was successful and Bill received a substantial sum in back pay for the period of the appeals procedure, together with a weekly pension. Sadly, Bill died soon afterwards but he was happy in the knowledge that he had won his case and that his caring relatives could share in his compensation money. This case strongly illustrates the argument that, where applications are at first rejected, but the Society feels that there is a reasonable case for a compensation claim, support and effort should not be abandoned.

Frederick was born on the South Coast and he joined the TA in 1935 as a driver. By October 1939 he was in France with the British Expeditionary Force (BEF). When his convoy was bombed, en route for Dunkirk, he hid under his truck and was knocked unconscious by a near direct hit. Fred has no idea how he got back to England. The Army kept him in the Service but Frederick found that he could not keep his balance. When he was eventually put on to driving a staff car, he began to experience panic attacks. Frederick tried in vain to master his fears until he collapsed in the street and was taken to Mill Hill Military Psychiatric Hospital, from where he was discharged in 1942.

Frederick worked at first as a labourer repairing bomb damage and he was granted a War Pension. He tried another job as a driver then he became a milkman, but he lost too much time through illness. He was persistent at every job he tackled but every time there was any redundancy Frederick was always the first to leave because of his sickness record.

He last worked twenty-five years ago and now he hardly ever goes out. He is bitter about not being able to work. This made him depressed, and he is now agoraphobic. He is afraid of sudden loud noises and has a long suffering wife. The Society got him into a private nursing home locally to give her a much-needed break, but it only lasted a few days before Frederick was insisting on being taken home. He now flatly refuses to go anywhere. Again, military psychiatric help would seem to be needed followed by a period in a military convalescent home. Because of his unco-operative behaviour, most of the other ex-Service organisations now feel unable to accept him.

I found **Alf** living in utter squalor in an upstairs flat in London, although he has 100 per cent War Pension for injuries sustained at Dunkirk. He received a full medical discharge in 1943. Alf was paying a substantial rent with no Housing Benefit rebate because of his maximum pension; his landlord was also profiting from Alf's payments into the electricity meter. Alf was unable to sustain a reasonable conversation as he had a very poor memory. He had no family or friends. In the winter he went to the local pub for warmth in the winter, even though he was not a drinker. I was able to support his case for Local Authority sheltered housing where someone might keep an eye on him. We never did succeed in getting Alf to come to Tyrwhitt House. He always had an excuse, or he simply just didn't turn up.

Albert went to France as a regular soldier in 1939. He was captured at Dunkirk and spent five years in Germany as a prisoner of war, where he was badly treated. However, he stayed in the Army and went to Korea in 1950. But it all became too much for him and he was discharged in 1955 with a 100 per cent War Pension for psychoneurosis. His was a case of an accumulation of traumatic experiences over many years. He gave up any form of work in his fifties and, following the death of his wife, he has benefitted enormously from visits to Tyrwhitt House.

Yet another Dunkirk veteran was **Owen** who escaped from France by swimming out to a destroyer. He worked on amphibious vehicles and went to Normandy in 1944. George crossed the Rhine on one of his own vehicles and all was going well until he witnessed the end of the appalling atrocities at Belsen Concentration Camp. George's military Service ended in various military hospitals. His small War Pension stopped when his family employed him in their business, and it was felt, at that time, that he had recovered. When I met George for the first time he was receiving psychiatric counselling and I was able to initiate a case for the reinstatement of his pension, besides offering him a break at Leatherhead. He still goes there and the Society has arranged that his pension be restored to him.

*   *   *

Britain was involved in the defence of Norway in 1940. I first met **Graham**, a retired Royal Navy Lieutenant Commander, in 1985, in a small village in the West Country. He told me he ran a few sheep and kept a few horses in the paddock near his rambling and antiquated house where he and his wife lived. Graham was permanently on edge and his mood was liable to change suddenly. He could be verbally offensive one minute, and charming in his next sentence. Much of what he said was utter nonsense and his wife used to go away for a few days to get away from him. Graham would then fend for himself with food left in the freezer. He showed me a letter from his last employer indicating that they would never even consider taking him on again.

Throughout most of the war Graham was employed on shore duties. He was a pre-war Dartmouth cadet who saw many of his colleagues become very senior naval officers. After the war he applied for discharge with the 'golden handshake' many others were receiving but the Navy refused to let him leave. Many years afterwards, Graham was permitted to retire from the Service and, after many appeals, he received a 20 per cent War Pension for mental illness aggravated by the Service. He was diagnosed as schizophrenic. At an interview with the Society's Chief Consultant, it was felt that his pension was totally inadequate, but that Graham was completely unsuitable for admission to any of the Society's homes because he would be likely to upset and distress the others.

I eventually took him to see another, independent, consultant, at the request of the War Pensions Board and, at yet another Appeals Tribunal, it was revealed that, at the age of seventeen, as a Midshipman, he was blown from one side of the ship's bridge to the other whilst under enemy attack in Narvik Fjord, in May 1940. A senior officer present remembered that Graham joined the Navy at the age of fourteen, in perfectly good health. The Service accepted responsibility for his ruined life and its consequences.

Graham came away with an 80 per cent War Pension, attributable to the Service, together with substantial arrears of pay covering all the years of his disallowed appeals. Very sadly, he did not live to enjoy his good fortune for more than a few months. In this case perseverance paid off, but I feel that the excitement and strain of it all were too much for him.

Quite a different story was that of **Raymond**, who joined the Royal Air Force at the age of eighteen in the 1920s. He was commissioned as a fighter pilot and, like Graham, saw Service off Norway in Spring 1940. Despite receiving head injuries he still managed to fly his plane back to base in Scotland. For this and other brave deeds Raymond was awarded the Distinguished Flying Cross (DFC).

After an understandably lengthy period of recovery, Graham went to the United States as a flying instructor. He returned to active Service flying Hurricanes over Burma. He later held several senior Air Force appointments after the War and eventually retired as Group Captain. I met Graham in a private nursing home after he had had a stroke. He was reported to me as being a difficult patient with unsociable behaviour. This was hardly surprising, because Graham's life had been totally unlike that of the other patients and he was having difficulty in communicating with any of them. In his otherwise successful life he had had two disappointments in marriage.

We were able to take Graham to Tyrwhitt House where he thoroughly enjoyed himself with the other ex-Servicemen. He did not have much to say, and indeed could hardly speak, but he joined in the various activities and returned to his nursing home a much happier man, more easy to get on with. We repeated this exercise until Graham died.

Surprisingly, I never had an RAF Battle of Britain pilot as a client although many of them were known to the Society. My Battle of Britain casualties were from the RAF ground personnel, including an ex-corporal who was to become my last case, in April 1994.

Harold was totally incapable of fending for himself and had relied on his wife and daughter for the past seventeen years. Aged eighteen, Harold was arming-up Spitfires at RAF Hornchurch on 31 August 1940. No air-raid warning was given when some sixty bombs fell in his immediate vicinity. Six of his comrades, in a truck, were killed outright. On the same day, in the afternoon, the bombers returned and Harold saw one of his colleagues decapitated by a concrete block which fell on him. Harold was thought to have been in the truck and was reported as 'killed in action'. Badly shaken, he went to Sick Parade but they were far too busy dealing with the physically injured and the dying to attend to his state of shock.

In 1975 Harold visited the RAF Museum at Hendon, where they were shown some rare archive films of the Germans bombing RAF airfields. Harold began to feel quite ill with a severe headache and sickness; by the time a friend got him home, he was delirious. He cried and crawled on the floor and his speech was incoherent. Later that night Harold was screaming and shouting that they were under enemy attack from the air and that he was going to be killed.

Harold remembers very little of this but his wife will never forget. He had to give up his job as he was by now regarded as mentally unstable. At my request seventeen years later, Harold was seen by a military psychiatrist but the application for a War Pension was rejected. The argument was that he had continued to serve in the RAF for a number of years after 1940. We mounted a further appeal and an independent psychiatrist, a civilian, and I saw him again. Post Traumatic Stress Disorder (PTSD) was diagnosed and this led to the award of a pension. At the interview Harold was friendly and cheerful, but his memory was very poor and there was a total lack of any constructive thought from him.

Ernest Walker[1] lives in a bungalow in a country village. When I met him he was 88 and lived alone. He was a pre-war Corporal on the Ground Staff of RAF Maleme, which was abandoned when German paratroops invaded Crete and all British aircraft were withdrawn. The trench in which they were sheltering began to cave in under shellfire. After surrendering, they were falsely accused of killing wounded German soldiers and taken to be shot, instead of being treated as Prisoners of War under the Geneva Convention. One of the party spoke out, insisting on the presence of a German officer at the shooting. However, no commissioned German officer could be found at the time and, after a long period of uncertainty, Corporal Walker was incarcerated in difficult conditions before being taken by sea to Salonika two months later in a half-starved condition. He saw his best friend die after a long march in the middle of the night, in the rain.

Ernest spent the rest of the war in prison camps and later, as a civil servant, he was obliged to retire early through ill health. He only received a War Pension long after his eightieth birthday and the Society is appealing for an increase beyond his current 30 per cent. He has always experienced flashbacks to his wartime experiences. Writing a book about them was perhaps a useful piece of therapy. It certainly makes interesting reading about one of the lesser known events of the Second World War – the fall of Crete.

Charles was with the Eighth Army, in the Royal Artillery, in the Western Desert from the early days of the war. In the spring of 1941 he was transferred to Crete where he was knocked down by

a truck and received serious head injuries as the German para-
troopers attacked. He was fortunate to be evacuated to Egypt
before the island fell in May 1941.

Charles was discharged from the Army before the end of the
war but he was only able to do unskilled labouring jobs. One day
he blacked out and fell off his ladder. His War Pension was only
awarded in 1973 when it was found that his Service had led to
psychoneurosis, head injuries, sandfly fever and bilateral hearing
loss. He is now a regular visitor to Tyrwhitt House.

*   *   *

As the Germans pushed the Eighth Army back towards Egypt,
having captured Greece and Crete, Malta stood out in an Axis-
dominated Central Mediterranean area. While the island was
being bombed incessantly, the Army's attitude was that because
there was no escape from the island there would be nowhere for
cowards and deserters to run to. There was therefore no need for
psychiatrists! It was estimated that 25 per cent of the troops were
affected by the continual bombing. Evacuation of casualties was
quite impossible and there were many who were near to cracking
up completely. It was mistakenly thought by the authorities that
courage and cowardice were merely alternative forms of behaviour. Notices appeared telling the men not to succumb to fear – as
if most of them had any choice in the matter.

Imagine my surprise when I found that **Neale** had lived at
Milner House and worked at Thermega from 1947 to 1950. He was
a photographer's assistant before the War and on being called up,
he joined Air Reconnaissance. The RAF posted him to Malta
where the pressures of the work, its importance to the war effort,
and living with danger every time the enemy targeted airfields,
made Neale feel ill. After the seige was lifted, he was posted to an
airfield in the UK where he had a complete mental breakdown.
He was discharged from Belmont Military Psychiatric Hospital at
the end of the War and returned to his parents.

On leaving Milner, Neale stayed at home until his father died,
after which he got a job in a hospital X-ray department for ten
years. After his mother died Neale remained at home alone in
the family house, where he has been ever since. He never received
a War Pension, nor would he accept further help from the

Society. Neale always welcomed the occasional visit from me.

Paul lives in Devon. He had driven a taxi before the war. The RAF trained him as a signals wireless operator in the Ground Staff. Like Neale, Paul suffered extreme distress during the seige of Malta but somehow managed to carry out his duties, albeit with great difficulty. Probably because of this he never succeeded in getting a War Pension. Simply because he had managed to survive, the authorities felt that Paul had suffered no harm.

On returning home, Paul rejoined the family taxi business but was unable to cope with the stress of driving. Paul had had a good education and later was able to start up an antiques business in a seaside town. Luckily, he had an assistant who was able to stand in for him on the many days he was unable to concentrate and face the customers. Now retired, Paul's GP says he suffers from an anxiety neurosis. He cannot always manage to go out of doors and relies heavily on his caring wife.

The Society is hoping to persuade Paul to come to Tyrwhitt House where the case for a disability pension might be investigated. He certainly deserves our help.

*   *   *

The remaining aspects of the War to be discussed in this chapter are the events which commenced in North Africa. First, there were the deeds of the Eighth Army under General Wavell who all but drove the Italians to complete defeat until forced back to Egypt by the German Afrika Korps; second, the counter-attack from El Alamein through Tripolitania; and third, the situation with the First Army which landed in Algeria, with the Americans, and joined up with the Eighth Army in Tunisia. The Allies then invaded Italy through Sicily, until meeting heavy German resistance at Monte Cassino and across to the Adriatic. It was then February 1944.

For the men in the Middle East there was a variety of scenes, an unfamiliar environment, occasional battles interspersed with sand, flies and boredom. Psychiatric centres were successful in conserving manpower through rehabilitation processes. Adequate sedation was important, military discipline was maintained and there was contact with comrades, reinforcing individual and group morale. The success of these centres depended on early diagnosis and treatment while sometimes a posting to a new job might aid recovery.

By the time they reached Cassino, where the Allied advance into Italy was halted, extreme battle conditions prevailed. The Americans concluded that the average GI (General Infantryman) was finished as a fighting soldier after an aggregate of 200 to 240 days in combat. They said that a man reached his peak at ninety days, then his efficiency wore off. Many who survived Cassino were never the same men again and could not be successfully returned to the front line. The British lasted longer in battle than the Americans because they were pulled out of the front line for four days' break after twelve days in action. On the other hand, the Americans might be expected to fight for up to thirty days at a stretch.

In the Middle East theatre of war the main problems were those of distance for bringing up supplies and the evacuation of casualties. Many of the soldiers there should never have been brought to the battlefield. After breaking down, some soldiers were found to be schizophrenic. Sixty per cent of all Middle East psychiatric casualties never took part in a battle at all. During the retreat to Alamein, May to July 1942, things went badly wrong for the Eighth Army as the men became demoralised by rough living conditions and shortages of equipment. The men felt unsafe.

After the Battle of El Alamein, and much morale building by General Montgomery, 60 per cent of casualties with slight problems of anxiety, taken to Rest Centres, were back on duty within a week. When the Eighth Army reached Tripoli in 1943 it was found that 97 per cent of the psychiatric casualties had had some form of history of instability. In spite of this, however, the Advanced Psychiatric Unit returned 93 per cent of these to some form of duty within a period of twenty-eight days, after rest and sedation.

By the time the men reached Tunisia, it was a more dangerous terrain. The Germans were determined to cling to their last remaining hold on Africa and, in the Eighth Army, exhaustion set in. It was considered that no officer should stay for more than a year in operations. As for the men, it was felt that they could remain under active Service conditions for perhaps two years before there was a loss of morale, interest and efficiency, which in turn could lead to delinquency.

The Central Mediterranean Force, which included the British First Army, had no psychiatric back up at first. Moreover, as they advanced eastwards, the lines of communication stretched right back to Algiers. Many patients evacuated there arrived in poor shape. Only a few returned to fighting units but 75 per cent of all psychiatric casualties in Algeria went back to some kind of military duty. As the forces combined to invade Sicily, casualties were returned to Tripoli. By the time the Allies were experiencing difficult battle conditions in Italy in 1944, Corps Exhaustion Centres had been set up and the earlier evacuation of mental casualties was found to secure even better results than had been the case in North Africa.

Good morale had been recommended by the Southborough Committee in 1922 as being the most important object of military training. It was a vital weapon against desertion, as was that of identification with a group whether it be a company, a platoon or a regiment. Factors which might lead to desertion included waiting in a transit camp, separation from familiar company, and the death of a comrade or an officer. Feelings of personal guilt could build up, together with bad news from home, or the lack of any news at all. Desertion was highest in the Infantry where the work was regarded as unrewarding and unspectacular.[2]

Infantry soldiers had the lowest expectations of their prospects in civilian life after the war. Psychiatric breakdown often originated from poor leadership and lack of team spirit. No one realised this more than General Montgomery, who stated that whenever soldiers got into trouble it was nearly always the fault of some officer who had failed in his duty towards his men.

*   *   *

My clients who served in the North Africa and Italian campaigns came from a variety of backgrounds. The Eighth Army fought in the Western Desert using tanks and armoured cars. **Harry**, who lives on the South Coast, was a mechanic in military workshops. He gave me graphic accounts of receiving damaged tanks from the battlefield only to find dead bodies inside. This experience gave him nightmares after the war. Harry took part in the invasion of Sicily, working inside an area booby-trapped with mines. And, again, there were grisly sights inside the vehicles recovered for repair. Harry never received any help or compensation to cope with the consequences of these shocks, nevertheless he did manage to work after the war. Now that he is retired, the nightmares are more frequent. The Society makes regular visits to him.

Brian was a pre-war member of the Royal Armoured Corps. He was awarded a War Pension after discharge at the end of the war. This was stopped after six months because,he told me, he never kept in touch with the Ministry of Pensions. He described vividly how the armoured car he was driving blew up under fire in the desert. His officer was killed immediately and Brian was set on fire, with his hair and clothing alight. He struggled out of the vehicle and reached the nearest medical centre. The shock of this finished Brian as a soldier and he had a spell at Northfield before discharge in 1945. After a long period of unemployment he was eventually taken on as a builder's labourer. His wife has been the breadwinner since 1978. Brian has all the signs of PTSD, with facial pains and dental problems. He says group therapy does him no good and he refuses to leave his wife in order to come to Tyrwhitt House. Consequently, both he and his wife continue to suffer from his nightmares. We were recently able to get his pension restored and we are hopeful of giving Brian and his wife a Not Forgotten Association holiday in Bournemouth.

David has unpleasant memories of the Eighth Army. He was a pre-war soldier who was already in the Middle East in 1939. He was blown up by a bomb at the seige of Tobruk. David became depressed and, later, he went absent without leave in Cairo. His days in detention were described in some detail, including digging an enormous hole in the sand one day and then being told to fill it all in again on the following day!

David was eventually treated for depression at a psychiatric hospital in South Africa and was medically discharged. He had joined the Army with an obsessional dislike for his father who had abandoned the family; David had transferred this dislike for his father to other forms of authority, both in the Army and in later life. David's War Pension was for psychoneurosis 'aggravated by the Service' and any deterioration in his condition will be regarded officially as due to other factors, such as ageing. He last worked in his fifties and complained about this until I reminded him of the hundreds of cases where my patients had not worked at all since leaving the Services. David has been to Tyrwhitt House and I believe it did him good. He does not enjoy life and seldom goes out. Recently, he has been reluctant even to see me.

David has always been unwilling to explain his problems fully to the Society's consultants. He is a manic depressive who appears to wallow in his own self-pity and who actually enjoys being miserable. When he came to the Society's convalescent home it was his wife who benefitted most.

Bertram was a Gunner in the Royal Artillery in the Western Desert. After a poor work record and ill health for most of his life, the Society managed to help him obtain a War Pension, but this was only about a year before he died. When he was terminally ill in bed, Bertram's voice changed to that of a young man and he went through the full sequence of verbal commands for preparing his field gun for action, loading it and firing it. At the point of death, Bertram had returned to his artillery battery at Alamein, reliving the episode in his life which had most affected him ever since. Because Bertram died as a result of his wounds, both mental and physical, his widow was able to receive a share of his pension for the remainder of her life. As I helped Bertram's wife complete the application form, she showed me a recent letter from the War Pensions tribunal indicating that Bertram had been allowed an additional care allowance with arrears of many years. This turned out to be a considerable lump sum. However, her husband never knew about this.

The First Army landed with the Americans in Algeria in November 1942, encountering fierce resistance after the initial element of surprise. **Frank** joined up, aged seventeen, and remained in England until he landed at Algiers and saw active Service right up to the German capitulation at Bizerta in 1943. He was severely wounded outside Tunis, receiving injuries to his left arm, right leg and the left side of his head. He was invalided out from St. Peter's Hospital, Chertsey in 1944 and awarded a 30 per cent, attributable, pension, which has never changed.

Frank worked as a plumber for many years. However, he began to suffer from strokes and the last one left him unable to walk. I first met him in a wheelchair, in a geriatric hospital. He was far younger and more active than the other patients. I never succeeded in getting Frank considered for Tyrwhitt House or Kingswood Grange as neither home, in those days, could accept a resident in a wheelchair. That is not the case now, and wheelchairs are most welcome! In my opinion, there should have been an assessment as to whether Frank's history of strokes was related to his head injuries.

Peter was a very fit Physical Training instructor in the pre-war Regular Army. He was allowed to transfer to the Royal Electrical and Mechanical Engineers (REME) as a vehicle mechanic with the First Army in Algeria. The dangers and pressures of his work led to him suffering from battle fatigue no less than three times until they had to evacuate him from Algiers in 1943. He spent the rest of his military Service in hospitals, where he suffered from nightmares. Frank told me that they gave him electro-convulsive therapy (ECT) which did not help him. He received a small pension for nervous anxiety.

Peter was determined to earn a living and this he achieved by becoming a self-employed window cleaner so that he could work out of doors and at his own pace. However, there were long periods when he could not work at all, and his wife stood in for him. When she died, he looked after their two children single-handed. They are now grown up and Peter lives alone, trying to socialise whenever his physical disabilities will permit. Peter enjoys coming to Tyrwhitt House. The Society helped him improve his pension, and at a later stage it might be possible to find a place for him at Kingswood Grange.

Edward volunteered in 1939. He was invalided out in 1946 after almost continual active Service. He went to Malta in 1940, thence to Tunisia before being sent home to prepare for D-Day. He landed in Normandy, and fought right through France and Belgium to Holland. It is not surprising that he ended his military Service at Northfield, where he was awarded a 100 per cent War Pension for psychoneurosis. Edward had been in the psychiatric hospital where I found him for over forty years, ever since leaving the Army! I found him friendly and sensible, a good example of a long term psychiatric in-patient, who might possibly be capable of rehabilitation in a home together with other disabled ex-

Servicemen. As a high percentage War Pensioner, financially he will be well provided for wherever he goes. As Edward has been in an institution for nearly all his life, it would be wrong to apply the recent 'Care in the Community' philosophy too literally, leaving him short of the care which he will certainly need and on which he has depended. In a similar case, admission to Kingswood Grange was unsuccessful.

A large number of my clients ended their active military careers at Monte Cassino or Anzio, where the Germans halted the Allied advance into Italy during the winter and spring of 1943/44. **Patrick** was commissioned into a Scottish regiment. He left the UK for Algiers in June 1943, and fought through Tunisia, Brindisi, the Salerno landings and Anzio. In May 1944 he was badly wounded on the Anzio beach-head when a bullet went right through his arm, and he was thought to be dead. One of his men rescued Patrick out of the line of fire and was subsequently awarded the Military Medal for gallantry in the field. Patrick spent the next two years in hospital before being medically discharged. He later received a small pension for his physical injuries.

In the 1950s, Patrick began to have psychological problems. He believed that he was being followed by enemy agents who were spying on his every move. His marriage did not last as his behaviour became obsessive. He managed to work until he was sixty years old and, when I met him, he was living alone and continually apologised for the state of his untidy flat. Patrick needs friends and support but he is a proud man who steadfastly refuses to accept charitable help. He is very much afraid of being regarded as a mental case. I believe his regiment may be able to give him some social life but he badly needs military psychiatric counselling. Patrick feels that to accept any help from the Society would be tantamount to admitting that he was 'mental'. We have been in a state of impasse with Patrick for many years.

Gerald was born in Southern Ireland. He joined the Army straight from being a farm labourer at seventeen and a half. He saw pre-war Service in India and completed his Regular Army engagement in 1938. He got a job building air-raid shelters and on being recalled to the Colours he was with the first British battalion to go to France in 1939. Gerald escaped from Dunkirk, went to

Algeria with the First Army, and thence to Italy where he became shell-shocked in 1944 after prolonged active Service. He was kept in hospital there until the end of the war when he was given a medical discharge and, later on, an 80 per cent War Pension for psychoneurosis.

Gerald returned to his sister and brother-in-law in London and has never worked since. He was put into a mental hospital whenever his behaviour became uncontrollable but Mary, his sister, always brought him home again when he quietened down. Since her husband died, the two have lived together even though she had recently suffered a stroke when I first met them. We tried very hard to take Gerald to Tyrwhitt House to give Mary a break, but Gerald refused to go anywhere for fear of being kept in. Mary kept him smartly dressed and Gerald used to do the shopping. He also cleaned the house, under supervision. I found them a gentle, pleasant couple with few friends. When I first met them they had never claimed any attendance allowance. Mary is almost ninety and, one day, I expect Gerald will end up in Kingswood Grange where he will certainly be most welcome.

Roger was an insurance clerk before the war, and a lance-corporal in the Territorial Army. When war broke out he was commissioned and went to Egypt as a Staff Captain with the Eighth Army. Roger landed at Salerno where he was wounded and saw several of his men killed under enemy fire. In August 1944, at Cassino, he received a further head wound but he kept going for several days after being further wounded in the back by mortar fire as he and his men held on to an empty house. As other officers were killed, Roger received a promotion on the battle-field, but he felt he did not deserve it. The fighting became so fierce that on one occasion his Sergeant Major had to fire his pistol in the air in order to discourage the men from retreating. We agreed that this was reminiscent of the First World War.

Roger was back in the UK by Christmas 1944, after three months in hospital in Italy. On medical discharge he received his War Pension, the amount of which has fluctuated. Roger took a variety of executive positions which came to naught because of his lack of confidence. He tried to start up his own business, but this venture also failed. He suffered nervous breakdowns and feelings of insecurity. Roger still suffers from severe headaches and his caring wife has had to go out to work in order to bring up their two children. A stay at Tyrwhitt House gives them both a well-deserved break.

As a pre-war territorial, **Tony** was embodied into the Regular Army on 2 September 1939. He was medically discharged in 1946. Tony went through Dunkirk, El Alamein and Sicily. At Cassino, he was defending a building which received a direct hit – an incident from which he has remained shell-shocked ever since. After the end of the war Tony received a 20 per cent pension for psychoneurosis. He became a Post Office telephonist but had to have spells off work for bouts of nervous anxiety. As a result, he was always passed over for promotion. Although he was well qualified, he could not make decisions. The jaundice he contracted in Egypt has recurred ever since. Tony recalls an incident before Dunkirk concerning a decision he took which resulted in the death of some of his comrades. His survival has left him with feelings of guilt, which have persisted ever since.

Tony and his caring wife were married just before he sailed for North Africa in 1941. They live quietly in their own house. Tony has been a regular visitor to Tyrwhitt House for many years and this has been of tremendous benefit to both of them. His pension has been increased. Although the Society has helped Tony and his wife, the problems caused by his wartime experiences have never left him. Even the food he eats is conditioned by his jaundiced liver.

Harry Byrne was a Gunner in the Royal Artillery. He was in the Battle of Ortona on the Italian Adriatic coast in 1944. As a member of a Forward Observation Party, Harry was often ahead of the British lines from where they reported back the effectiveness of our artillery fire besides actually directing the shots towards the enemy. "Observers" were liable to receive attention from the enemy and from our own guns, by accident. The worst situation arose when the Germans bombarded the house Harry was in with a very heavy gun mounted on a railway truck. The house took a direct hit and Harry began to suffer from shell-shock, although he continued his duties for the next four months. This was significant because the authorities later insisted that he had recovered from this incident when he applied for a pension.

At the end of the four months Harry's feet began to sweat and

Harry Byrne and his family, just after the war, and more recently.

bleed, a condition somehow connected with his nervous state. In 1947, he was awarded a 6–14 per cent pension for hyperhydrosis (severe perspiration) which lasted for a year, after which it was withdrawn. Harry returned to his regiment for a further seven years before receiving a medical discharge in 1954, suffering from hysteria.

Harry's career since leaving the Forces has been poor. He did clerical work such as book-keeping but he began to experience panic on buses, became afraid to go out of doors and, when returning from work, found himself clinging to railings and hiding behind walls. He has not been out of his house for the past ten years.

I suspected that these problems were aspects of Harry's feelings of a need to always take cover from an anticipated enemy attack from somewhere or other. I was sure that to deny him a pension was wrong, as without one Harry and his wife received only the minimum Social Security allowances. As part of a lengthy process of appeals, Harry was visited at home by consultants, and at first there were some disappointments. Nevertheless, they did get a television from the Not Forgotten Association, which helped to show them that somebody cared. The final breakthrough occurred in 1993 with the award of a 50 per cent War Pension. It was accepted that Harry's agoraphobia is, and has always been, attributable to his military Service.

Harry still sits indoors but life is now a little easier for him, and for his wife, who was able to afford a short holiday in Rome.

*   *   *

These case histories describe typical situations encountered by my clients. While every soldier's story is unique, there is one thread of similarity that runs through all the case histories. And that is, without help from the Society, and from family and caring neighbours, these old soldiers would have continued to experience their problems in isolation from the rest of society – the society they had fought to protect. It is sobering to think of just how many similar cases there still are today where the caring organisations, and neighbours, are totally unaware of the problem. If no one knows, no one can care, and no one can help.

Notes

[1] See Ernest Walker, *The Price of Surrender: 1941 The War in Crete* (Blandford, 1992).

[2] The PBI, or Poor Bloody Infantry, was a term which came from the First World War in France.

4

The Second World War: Casualties of the Far East and of Europe

Burma had been promised independence before the war, and it was the general perception at the time that the Japanese would never regard the country as worth occupying. The terrain was mountainous, with rivers and dense jungle. In 1941 Britain had two incomplete battalions of infantry stationed there, with weak air support.

When the Japanese invaded from Thailand in January 1942, with strong air cover, there began the long retreat which lasted until May. In 1943 Wingate's Chindit expeditions achieved the aim of tying down the Japanese and disrupting their communications, but Burma remained in Japanese hands until towards the end of hostilities. While many Burmese were loyal to the British, the Japanese Army was accompanied by individuals trained to be nationalist spies, and they stirred up the local population against us.

In five and a half months the British Army retreated one thousand miles and suffered ten thousand casualties. The men experienced physical and mental exhaustion caused by climate, illness and intermittent fighting with an enemy well trained for jungle warfare. The periods of heavy monsoon rain did, however, allow for some psychiatric treatment in the forward areas, with good results. Some soldiers were affected by lack of food, lack of jungle training and having to abandon casualties (6,000 were thought to be "missing" in Burma). General Wingate thought that toughness in a soldier was far more important than the prevention of illness and he was proved wrong. F. Spencer Chapman wrote: 'It is the attitude of mind that determines whether you go under or survive – the jungle is neutral'.[1]

Casualties arose from malaria, dysentery and nervous conditions, and Wingate's views were to be modified after he caught typhoid. Field Marshall Slim thought otherwise and gave the men reassurance; eventually psychiatrists, working with drugs brought to the front line on the backs of mules, were able to return 70 per cent of all psychiatric cases to active duty.

To any soldier, unprepared, the jungle was simply frightening, with utter darkness at night, strange noises, a hostile terrain with swamps and unclimbable mountains. The enemy were well trained and one never knew whether the natives were friendly or not. The majority of my clients who fought in Burma were subsequently able to live near normal lives immediately after the war. Some stayed in the Army, and some were the victims of adverse circumstances afterwards. But they all experienced recurring nightmares in later life, and few were ever able to work to normal retirement age. The jungle took its toll of them in the end. Only the highly exceptional came out unscathed while the most serious casualties never returned at all. They were among the six thousand soldiers reported missing, and never found.

Derek was called up in 1941 straight from school. He was commissioned into the Royal Engineers in India and for two and a half years operated behind enemy lines in Burma, laying mines and blowing up bridges. One day he left his camp in the jungle to pick up supplies which they knew had been dropped by air. Before he was able to return to camp, his own men opened fire on his party, thinking they were a Japanese patrol. He was pinned down all night and had to remain in a trench surrounded by his Sikh soldiers who had been killed all around him. Derek went five nights without sleep before rejoining his group. Later, he was ambushed by the Japanese but managed to escape.

Derek stayed in the Army until 1947. He then went to university, but could not withstand the learning pressures of the course. Ten years later he was able to complete his degree, part time, whilst working in a factory. He was then able to teach in a school but had to give it up after only ten years. Derek's marriage failed and at the end of his working life he was only able to teach a few individual students. I found Derek on his own, in a flat, a sad and lonely man. I encouraged him to apply for a War Pension and he was successful. A shy man, Derek desperately needs friends.

Nicolas lost his wife in 1988. He now lives with his unmarried son in a house full of motor-bike parts and general chaos. He was clearly very pleased to meet someone of his own generation and he talked incessantly. Frank never qualified for anything like a full State Retirement Pension because he had had long spells in a psychiatric hospital and he gave up trying to work when he was in his fifties.

Nicolas was with a Field Ambulance in Rangoon in 1942 and walked through the jungle for the whole of the thousand-mile retreat to Imphal in India, suffering from malaria. He recovered and somehow kept going through Iraq, Egypt and the fighting in the Po Valley in Italy. He had reached Austria by the end of the war. I have met other cases where men in the RAMC struggled to remain on duty in spite of illness, trying to set a good example to the others, especially when the Corps was short of trained personnel and casualties were high. Nicolas survived the war in this way until he was posted on the staff of Netley Psychiatric Hospital, where he broke down in 1946. He was discharged, as a patient, from Northfield and awarded a 50 per cent War Pension for psychoneurosis aggravated by the Service.

When Nicolas obtained work as a labourer his pension was reduced to 40 per cent as it was felt that his disability had lessened. He now manages on his small old-age pension with invalidity benefit and his War Pension. The Society arranged for his War Pension to be reconsidered. As a rule, it is not easy to have a pension increased for an aggravated condition because it can be argued that any deterioration could be brought about by other factors, such as old age.

Ned was a light infantryman based in India who was wounded on a Chindit expedition behind Japanese lines in Burma. After the war Ned did labouring jobs but he was not able to cope with factory work. He was periodically treated for depression, and continued to receive treatment for the malaria he first suffered in India. After his wife lost her job with the School Meals Service, they were desperately short of money. They had never asked for charitable help of any kind. Ned was referred to me by his GP who felt that the Army had contributed to his health problems.

When I discussed the case for applying for a War Pension, Ned and his wife were very reluctant to proceed with the application. Neither of them would consider him having a short stay in a convalescent home in order to have his condition assessed. They were suspicious of other people prying into their private affairs; he was afraid of being 'put away' or given treatment of which he might not approve. Several visits later I persuaded Ned to see the Society's Chief Consultant on a one-day visit with door-to-door transport supplied. As a result of this visit I helped him complete the forms and, with our Consultant's support, we obtained a small percentage War Pension. However, the story does not end there. When the Local Authority were told of their small improvement in weekly income, the rent for their Council house was increased accordingly as they were now deemed to be better off. Most local authorities disregard war pensions when assessing a couple's income for rent assessment purposes. But Ned and his wife happened to live in a place where the award of a War Pension to a disabled ex-Serviceman is considered in the same way as wages earned by a fit person. War Pensions are not free gifts; they are compensation from a grateful country for injuries sustained in the Service of our country. They are designed to make up for the fact that the average pensioner will not be able to earn as much as if he had not been injured at all.

Ned and his wife were furious. And from then onwards they have refused to accept any further help from me at all. I had hoped that an increase in his pension might be a nett gain for them, but they would not be convinced. I think they thought that because Ned's pension had led to an increase in their rent, I was some sort of spy for the District Council!

*   *   *

In the Far East, the dreadful fate of those who were captured by the Japanese has been documented in books and films. In 1942, a survey made by Army psychiatrists about prisoners who had escaped from Europe found many unable to rehabilitate themselves. Those who returned to duty had breakdowns and even the regulars needed sympathetic treatment. By 1945 doctors were warning of difficulties for war prisoners attempting to integrate into the community. It was pointed out that many would not show any symptoms for several months.

Some ex-Servicemen found old feelings reawakened by wartime films. There were also problems experienced by those who witnessed the concentration camps atrocities at the time of liberation. Some were to be deeply affected long afterwards.

If the German camps were scenes of acts of perversion, the Japanese camps were at least true to their own harsh culture in which their own soldiers could be thrashed, tortured and put to death. Their own prisoners were treated as outcasts when they returned to Japan in 1945. The surrender of 130,000 Allied Servicemen to 30,000 Japanese was unthinkable to them. They would have fought to the last man and no soldier was of any use if he would not or could not fight; if a Japanese soldier was killed he was a hero. Because of this, our prisoners were regarded as cowards. It was only the fact of being part of a large collectivity of prisoners that saved their lives. Patients in hospital when Hong Kong fell were stabbed in their beds. On the Burma railway one prisoner died for every seventh sleeper laid. All labourers were expendable and on one bridge alone 600 lives were lost. But if any Serviceman tried to escape, not only was he shot but so also were his Japanese guards. Pilots who bombed Japan, if captured, were executed as a matter of course.

In 1945, thousands of ex-prisoners came home with neither medals nor combat experience to give them any self-esteem. There were particular feelings of bitterness towards the Japanese. Many have remained silent about their ordeal. When they were released, many did not admit to their injuries because they wanted to get home as quickly as possible and not want to be detained in a military hospital for diagnosis and treatment. They were given double rations of food for six weeks and after that they were on their own as far as the authorities were concerned.

A survey of American Second World War prisoners in Japan revealed that, forty years on, 82 per cent were mentally impaired, 60 per cent had anxiety disorders, 18 per cent were depressed, and 28 per cent had PTSD. Many of them had more than one of these conditions.

Bob was my first Far East Prisoner of War (FEPOW). I was his welfare officer and I remember our first meeting, ten years ago, as if it were only yesterday. He lived in a tiny room, rather like a prison cell, with one window which faced onto a brick wall.

Within, everything was neatly arranged, clean and tidy. The kitchen corner resembled a ship's galley and the bed, with blankets neatly folded, resembled a barrack room awaiting inspection. It was January 1985 and I said, 'Good morning. Did you have a nice Christmas?' 'Don't talk to me about Christmas. I always go to bed on that day, the day we were captured by the Japanese, on Christmas Day 1941, in a frigate off Hong Kong. Do you run a Japanese car?' 'No I don't.' 'I wouldn't buy anything from those bastards after the way we were treated. Do you know the first thing they did was to line us all up on deck and then they killed all our officers in cold blood. It makes me feel sick every time I think about it.'

Bob then took out his War Pension allowance book and threw it on the floor towards me. 'Do you believe in God?' I replied that I did. 'I'm supposed to be a Catholic but I can't say I believe in God any more after what He allowed to happen to us. Here, take all my pension and make me feel better again.' At this point Bob knelt on the floor in front of me and cried his heart out. Words simply cannot describe how I felt. Bob went on to speak of despair, suicide and the heavy drinking he had indulged in as a means of escape from his memories. I felt the best I could do was to offer practical help rather than attempt to moralise with him. Bob was not short of money, provided he did not drink to excess.

I told him we must try to get him into somewhere more pleasant and where his daughter, then in her early twenties, might be able to stay, perhaps live, with her Dad. My predecessor had taken Bob to Tyrwhitt House but he had not settled there; he caused distress to the others, and had to be taken home. Bob begged to be given another chance at Tyrwhitt House but the Society's Chief Consultant, who controlled all admissions, felt that it might be too great a risk, at that time. I therefore arranged a holiday for Bob, through the Not Forgotten Association, on his own, at the seaside. He thoroughly enjoyed the break from his tiny 'prison cell'.

When Bob's wife died over ten years previously his daughter, then aged nine, was taken into care and he was completely alone. It was no wonder that his work as a painter and decorator suffered. Bob became depressed, he drank and the downward spiral of his life began.

Once we had helped him get a two-bedroomed flat from the

Council his daughter was able to stay with Bob, and later visit him occasionally after she got married. We helped him obtain a new carpet for the flat, which was pleasantly situated.

When I last saw Bob he had moved to an old people's bungalow where he saw his daughter from time to time. Significantly, photographs of his old ship appeared on the mantelpiece, and he was able to talk about his former days in the Royal Navy. I avoided mentioning the Japanese but certainly Bob was living in reasonable comfort and dignity. But he still had few friends. He was offered another Not Forgotten Association holiday but he felt there might be other, more deserving cases we ought to support.

Liam was referred to me by his GP's Practice Nurse who had discovered that he cried with distress at nights. On my first visit he told me how he had been hit on the head while working in a coal mine in Japan, but I suspected there was more to his problems than just this.

Liam told me how they sailed for Singapore and were landed in Java when Singapore fell, only to be taken prisoner there. He was a medical orderly in the prison camp sick bay, from where he tried to get clean blankets for his patients. Because of this, the Japanese accused him of stealing the blankets. They woke him in the middle of the night for interrogation and beating. To this day Liam wakes in the night screaming with terror. He imagines his bed is surrounded by hostile Japanese shouting and threatening him. Liam told me he had never spoken of this to anyone, not even to his wife. I was the first person to hear Liam's story in fifty years. He admitted feeling better for having related his fears to someone. Liam revealed that his GP had been called in when he became suicidal two years previously. He had suffered from periods of depression lasting several days ever since he came out of the Army at the end of the war.

After he was demobilised in 1946, apparently fit, he could not work for the first two years. He never applied for a War Pension and eventually went back to poultry farming, which he had done before he was called up. Whenever he had bad days his wife was able to keep the business going, collecting the eggs and selling them. Liam gave up work at 62 when he became too ill. Prior to this, he had had a breakdown and for two years, in his forties, he had been unable to work at all. This was one of the most serious cases I ever saw of an ex-prisoner badly affected by the events of fifty years ago.

I encouraged Liam to apply, and six months later he received a 40 per cent War Pension for PTSD. There was also about £1,000 due to him in back pay. When he went to Tyrwhitt House, from which he and his wife both benefitted, he was in a single room because of his nightmares. But this raised problems when he was ill in the night, unless a nurse on her rounds happened to see him at his worst.

In 1940, **Ambrose** was a railway signalman when he was called up for the RAF. Six months later he was on the ground staff at Seletar Air Base, Singapore. Taken prisoner when the Japanese arrived he was put to work in Java, and then Sumatra, where he caught typhoid and beri-beri. Ambrose must have been at death's door as the Japanese regarded him as too ill to send to the Burma Railway. He was still ill at the end of the war when he was sent to hospital at Bangalore, India. Ambrose finally came home a year later.

Ambrose tried to work but none of his jobs lasted because of his requiring time off through illness. Eventually, one of the newspapers in Fleet Street gave him a job as a lift attendant. He continued that job until he retired at sixty. Even so, he had long periods away from work through sickness. Here was a good case where a sympathetic employer made it possible for a sick man to preserve some of his dignity. Ambrose and his wife now rely on their daughter who lives nearby. His memory is patchy and he receives treatment from one of the military hospitals. Ambrose was always pleased to receive a visit, although his speech is not always easy to understand. He is a regular visitor to Tyrwhitt House and both he and his wife are very appreciative of this. Ambrose has a good War Pension which he certainly deserves. Although their lives were ruined by the war they regard themselves as more fortunate than many others.

Joe signed on as a regular soldier in 1938. He was in Hong Kong in 1939, followed by Shanghai for twelve months. When the Japanese invaded Malaya, Joe was on the Thailand border and was taken prisoner. He worked on the Burma railway before going to the coal mines in Japan. By the end of the war he weighed only six and a half stone, yet he was discharged as medically fit.

On arriving home Joe was too ill to work for two years, suffering from dysentery. He then worked for his brother but there were periods in hospital before he qualified for a low percentage War Pension. Eventually he was taken on as a hospital porter, a job from which he was able to have some time off when he was ill. Joe still has nightmares about fighting in the jungle. He now helps the local branch of the Far East Prisoners of War Society. It is hoped that he will benefit from periodic visits to Tyrwhitt House.

Conscription affected everyone. **Henry** went to a public school before working in the West End as a trainee estate agent. He was twenty-five when he was called up. Two years later he had fought the Japanese all the way down the Malay Peninsula until, finally, he was wounded and taken prisoner. Henry was made to work on the Burma railway and his condition was so obviously poor at the end of the war that he was awarded a pension straight away. As a professional man, Henry went back to estate agency work, but he could not settle down. He went to university to study medicine but gave up after a year. He went into surveying but again it did not last, so he tried pig farming for a living. Finally, Henry ran his own small building firm with the help of his wife on whom he now relies very much. Mary made it possible for him to carry on in business when he felt too ill and she gave him the encouragement he needed. Fortunately, she is also a retired nursing sister and she is able to drive him to hospital when Henry becomes unwell. Both have benefitted from a stay for Henry at Tyrwhitt House.

Jim had a career in the Royal Navy as a boy in the 1930s, becoming a telegraphist. From the outbreak of war he was on North Atlantic convoys before sailing to Narvik on HMS *Isis*, which was badly damaged and had to be towed to Falmouth for repairs. Jim joined the *Prince of Wales* as a Petty Officer, from her commissioning and fitting out. They sailed to the Far East where she was sunk by the Japanese off Malaya in December 1941. Jim was one of the few who survived, reached the shore, and lived rough until captured by the Japanese and taken to Java in 1942. For the rest of the war he worked in coal mines in Japan as slave labour. He was discharged as medically fit in 1946.

By 1949 Jim had a War Pension for injuries to his back when the *Prince of Wales* went down, aggravated when he was beaten as a prisoner. To this was added malnutrition, and privation with associated nervous symptoms. In spite of all this, Jim successfully pulled himself together and became a senior officer in local government. However, he was in a run-down state when he had to take early retirement, and he always had recurring dreams about the Far East. He later developed pernicious anaemia and can now only walk short distances. Tyrwhitt House has been of benefit to him and has given his caring wife a much deserved break.

*   *   *

Prisoners of war in Europe were, on the whole, better treated than those in the Far East; after all, large numbers of Germans and Italians were held prisoner by the Allies. However, as conditions became worse for the civilian population in Europe, so conditions deteriorated for the prisoners. And those who tried to escape were severely punished. Then there were the long marches and the bitter weather in winter. As the Allies advanced the Germans moved our prisoners to avoid their release, probably to use in some form of bargaining later on. There were nightmare journeys, packed into cattle trucks, where men, already undernourished, died of exhaustion. Ernest Walker describes such a journey from Greece to Silesia.[2]

Charlie joined the Territorial Army in spring 1938 and was quickly sent to France with the BEF in 1939. He was captured by the Germans before he was able to reach Dunkirk. Charlie spent the whole of the War in Poland. He escaped twice and was punished as a result. On the second occasion, he was recaptured by the Gestapo and put in a forced labour gang. The third time he escaped he reached Warsaw, where he was picked up by the Jewish underground resistance movement. Charlie remained there, in hiding, until the end of the war. Charlie had a successful career as a clerk with London Transport but, as he grew older, he suffered from heart attacks and blackouts until forced to retire early. His nervous condition was later aggravated by a virus, making it difficult for him to move and inflaming his muscles. We shall never know the extent to which Charlie's later mental and physical problems might be traced back to his five years, under great stress, enduring privation and harsh treatment in Poland.

Whenever I met Charlie, and until he died, he always explained that he and his wife were completely devoted to each other and quite determined to see out their lives together with the minimum of help from anyone else. He always insisted that, if we had a vacancy for him at Tyrwhitt House, it should be given to someone more deserving than he was, perhaps to someone who had not got a wife at all.

Geoffrey Vaughan is a client who lives in Devon and I visited him as he was completing his book *The Way it Really Was*.[3] In his introduction Geoffrey comments that most of the books about being a POW were written by commissioned officers. He wrote, 'I was an "other rank" and being captured meant something totally new in my experience. There were other reasons why I decided to write my experiences as a prisoner of war – personal reasons. My children and other members of my family who had only heard briefly about "my war" urged me to do so. I realised that a whole new generation had grown up since the war – a generation who could not appreciate that the friendly countries of today were capable of inflicting such hardships on their captives forty years ago.'

When captured, Geoffrey Vaughan describes how they were searched and anything of any value was stolen by the Italians, 'even wallets were taken, emptied, and the contents such as photographs and letters were flung contemptuously on the ground. The guards were also quite ferocious. I remember one Italian in particular was carrying a hand grenade gripping the ring between his teeth and looking quite ready to use it.' They were transported to Italy from Benghazi in the hold of a ship with the Italians throwing lumps of bread to them and laughing at the fights which broke out. 'By the time we docked in Brindisi we were a scruffy ragged sight, unwashed, unshaven, smelly, filthy, and in fact reduced to the lowest ebb.' They were to survive on Red Cross parcels.

The prisoners were eventually moved to Germany where Alsatian dogs were used by the guards. 'It is a fact that a dog was sent into the Russian compound one night and did not return. The Russians caught and ate it, and only the fur and bones were found (by the Germans) next morning.' At the end of the book, Geoffrey Vaughan concludes: 'It was a long time however, before I was able to eat a normal meal, and years before I regained the weight I had lost. There were also to be many bad dreams and, for a long time, I was always looking over my shoulder for the guard whom I felt should be there whenever I went out. But in the end things got back to normal – so much that I almost – but not quite – forgot the way it really was.'

Benjamin joined as a groom in the Royal Army Veterinary Corps, in 1939. He served in Palestine and Greece where he was captured by German parachutists in 1940 and taken to Austria. The interesting thing about this case is that Benjamin was a prisoner for five years and discharged after a further year suffering from psychoneurosis, but there was no pension.

Benjamin married, but became depressed. His wife left and he was in a mental hospital from 1950 until I met him in the 1980s. He spent most of the daytime outside the hospital, in the local town. Speculation on such cases is very tempting: was Benjamin one of those who should never have been accepted by the Forces in the first place? If his wife had stayed by him would he have settled down and led a normal life? He eventually received a War Pension, the authorities accepting that his sufferings as a POW had scarred him for life. When I met Benjamin, he had had no medication for some five years and was completely calm and friendly. Perhaps it was wrong to have kept Benjamin inside an institution for almost the whole of his life, and all at the expense of the public.

When war broke out, **Terry** had completed his first year at a teacher training college. He was in the TA and was soon posted to France. Terry was captured on 30 May 1940, as the Germans advanced towards the coast. He was held captive in France, Poland and Germany and was found to be suffering from malnutrition when repatriated five years later.

Although Terry was promoted to Sergeant on arrival back in the UK he was discharged in 1946 suffering from depression. He returned to college, qualified as a teacher of drama, and successfully followed this career in spite of periods of depression. Unfortunately, at the height of his profession Terry was forced to retire early and, later, a theatrical enterprise failed because he could not cope. He now does occasional play reading for charity. The root cause of all Terry's problems goes back to his days of captivity, and his extreme mood swings. He has feelings of guilt because he survived when many of his comrades did not. The abortive attempts to escape and his recapture still haunt him after so many years. I gave Terry advice on how to apply for a War Pension – something he had never done, for fear of reawakening bad memories.[4]

Tony lives with his wife in the Midlands. He only recently received a War Pension, for which he is grateful. His wife told me that throughout their post-war relationship his behaviour had been unpredictable and their relationship sorely tried.

Tony was not called up until 1943. He was captured at Anzio in Italy from where he was taken to work in the Polish coal mines. Although a prisoner for only thirteen months, he weighed only seven stones when rescued by the Americans in 1945. Many of his fellow prisoners died and he spent six months at an Army Rehabilitation Centre[5] after repatriation. He was eventually able to work as a welder but had to take early retirement.

Allan was awarded a 100 per cent War Pension in 1977 with the help of the Society. He contacted me asking for a visit to Tyrwhitt House for a break. He was obliged to retire from his work in the Civil Service at the age of forty-two. Later he tried to run a theatre box office. He suffered from severe personality upsets for most of his life. His speech was interspersed with long pauses, which he found were embarrassing.

Allan continues to be haunted about being a POW in Italy. He even tried to allay his ghost by returning there in 1973. After the Italian front collapsed in 1943, Allan, together with other prisoners, was sent to Czechoslovakia, where they were cruelly treated by the Germans. A final cause of Allan's distress, not caused by the war, was when his wife contracted multiple sclerosis and became physically disabled.

$$*   *   *$$

When considering prisoners of war as a group, certain common factors emerge. Many suffered a severe shock when taken prisoner – the experience was probably something which had not occurred could happen to them, or if it had occurred, the consequences could not have been imagined prior. Then there was no turning back. Subsequently, years later, there are feelings of guilt, of being restless and unsettled. They are unable to concentrate and moods fluctuate between extremes. These symptoms seem to occur regardless of the length of time in captivity or the quite different conditions in South-East Asia compared with Europe. Common factors with all prisoners were lack of food, forced labour, and, above all, the uncertainty as to what was to happen next, or who would be next to die from the effects of a cruel regime.

*   *   *

The largest number of battle casualties came from Normandy and the rest of the campaign which ended the war in north-west Europe. From July to September 1944, 20 per cent of all battle casualties were psychiatric, many suffering from mental exhaustion. For the first time in history, psychiatric casualties were anticipated. General psychiatric hospitals went to within ten miles of the front line. Only 10 to 20 per cent of their cases had to be evacuated to the UK and over the whole of the Normandy campaign. A third of all psychiatric cases were fully returned to duty (4,400 men).[6] One lesson learned was the screening of reinforcements, as it was found that 90 per cent of all exhaustion cases were originally in the wrong medical category.

Danny volunteered for the Territorial Army in 1939 and was sent to Iceland working on defence works in the Royal Engineers. He went to France shortly after D-Day, clearing minefields. He was hit by a shell near Caen and saw eight of his comrades killed. He was evacuated to hospital in the UK where he received treatment for multiple wounds until he was discharged in 1946.

Danny started up a small building business until he had a nervous breakdown in his fifties. His wife took over the business and he has stayed at home ever since. His War Pension was increased to include psychoneurosis as well as physical wounds, but he never went back to hospital. He felt he was mentally deteriorating and his wife wound up the business before I met them. Danny now benefits enormously from the reassurance he finds at Tyrwhitt House.

Hugh was a builder's labourer in South Wales until he was called up in 1942. In June 1944 he took part in the D-Day landings from which he survived unscathed as an infantryman. Two

months later, when his battalion encountered stiff German opposition as the British Forces swung out of Normandy, Hugh
received bullet wounds in the head. He was evacuated to a
hospital in Oxford, and finally discharged from the military
psychiatric hospital at Banstead, Surrey.

The Ministry of Pensions assessed him at 20 per cent, which
continued until I saw him in 1986. Hugh had managed to work as
a postman until he was fifty. He enjoyed the open-air life and was
known to everyone in the village. However the job became too
much for him, and he contracted asthma with severe headaches.
He looked after his widowed mother to whom he was very close.
The medication he was receiving for his headaches affected his
eyesight. Whilst at Tyrwhitt House, we were able to have him seen
by a consultant in London, who put him on a more agreeable
medication regime. His War Pension was subsequently recommended for a much deserved increase.

Jimmy studied to be an electrician whilst doing labouring work
before he was called up in 1941. He went to France on D-Day and
was blown up three times whilst riding a motor bike on the beach
as a despatch rider. He was left on the beach for dead until rescued
by some French Canadians. When he came round he began
waving his arms violently and had to be brought back to England
in a strait jacket. They put him in an asylum for three months
before transferring him to Belmont Psychiatric Hospital, where he
was detained for three years. When Jimmy finally arrived home
in 1949, he was unable to work at first because he continued to
have fits.

Jimmy eventually got a labouring job in 1952 and once he had
been taken on as an electrician with a lift company, the challenge
of the work kept him going, although he was still on substantial
medication. Jimmy has remained at seven stones in weight ever
since he left the Army. He finally retired at sixty. Jimmy's wife,
who is ex-ATS, is totally blind. Both of them are cared for by their
son who lives near and drives him to Tyrwhitt House to give them
all a break.

Dave was a trainee mechanic on the buses, called up in
September 1939. He survived Dunkirk, went to Iceland as a driver
in the RASC for two years, before returning to the UK in time for
D-Day. Dave was medically evacuated from the Battle of Caen

suffering from nervous exhaustion. His condition was such that he could not be returned to his duties and he was eventually discharged from the Army with acute anxiety state, for which he received a 30 per cent War Pension for an aggravated condition. The Authorities believed that Dave's condition was not caused by the war, but was merely made worse by it. At his Medical Board he had admitted to having had nightmares as a small child, before growing out of them.

Dave spent periods in the local psychiatric hospital, as well as receiving treatment for nervous anxiety and depression from his GP. Instead of being an engineer on the buses, all Dave could manage was to clean them. He married more than once and there were understandable difficulties with wives who could not cope with his illness. The Society was able to successfully assist Dave in having his pension reconsidered from 'aggravated' to 'attributable to military Service'. His percentage disability was increased to reflect more accurately his degree of disability.

Edgar lost his War Pension twenty years ago because the Electricity Board who employed him as a cleaner allowed him time off, with pay, when he was ill. Consequently he was never officially sick and the Ministry of Pensions reasonably assumed that Edgar had fully recovered from his war injuries. Edgar never thought to tell the War Pensions department what was happening.

Shortly after D-Day Edgar was found to be suffering from shell shock and was medically evacuated home. The Society assisted him in getting his pension restored.

Herbert served in the RASC from 1942 to 1946, when he was medically discharged for 'Failing to fulfil Army medical standards'. Herbert was the driver of an amphibious landing vehicle and many hours were spent before the invasion making mock attacks on British coastlines. The men would be soaked to the skin and often had to stay in their wet uniforms for hours on end. Herbert successfully landed his vehicle on D-Day, but not long afterwards he began to be very ill. He was taken to a hospital in Yorkshire suffering from tuberculosis and spondylitis of the neck and spine for which he received a small War Pension in 1991.

He managed to work up to the age of sixty-five as a self-employed mobile greengrocer going round the country villages.

As the years went by, takings fell as a result of competition from supermarkets. By the time he completed his last round, it was found that many of his customers patronised him because they felt sorry for a disabled ex-Serviceman. Consequently there was no goodwill to pass on to a successor and the business could not be sold to anyone.

Herbert now grows his own vegetables in his large garden, but he badly needs psychiatric counselling. He suffers from headaches and worries about the consequences of his Army experiences. He has a caring wife but needs to reduce the nervous tension which clearly goes right back to the war. The Society is arranging to have him seen by a military psychiatrist locally. His wife is far from well and Herbert is reluctant to be parted from her, even though a spell at Tyrwhitt House would undoubtedly do him good.

Victor worked in an aircraft factory until he was called up in 1944. He saw active Service in Belgium, Holland and the Ardennes in 1945, where the Germans made one last desperate attempt to halt the Allied advance. Victor received head wounds and was discharged from a military hospital with a 60 per cent War Pension for psychoneurosis attributable to the Service.

Victor has had a sad life, in and out of psychiatric hospitals. He was trained at a Government Training Centre to be a painter and decorator, but all his jobs were of short duration. He was convicted of manslaughter and served a long sentence in Broadmoor. One wonders to what extent his war Service was connected with this offence.

When I met Victor, he had been in a psychiatric hospital for the past four years, doing small jobs such as making beds in a geriatric ward. He has been conditionally discharged from the hospital, and the Society is assisting him in his rehabilitation. He now has a 100 per cent pension. Victor is obviously accustomed to a hospital routine and this will have to be considered in any plan to put him back into the community. Victor could be considered for Kingswood Grange, but the regime there is more informal than what he is accustomed to expect. Victor requires an environment of discreet but reliable supervision.

*   *   *

The war at sea consisted of a few full-scale naval engagements, some operations in support of our ground troops, but mainly escorting convoys across the Atlantic, the Mediterranean or round the North Cape of Norway to Russia. **Henryk Nowakowski** was born in Poland. Before the war he was a steward on the Polish-America Line which sailed from Gydinia to New York. He could speak nine European languages. When war broke out he was in Brazil on a Polish cargo ship. By December 1939, as they were

Rick Nowakowski when a Polish sailor and in 1996 at Tyrwhitt House with Dr Desmond Murphy. He is wearing the miniature of the War Medal presented to him by the Polish Ambassador.

sailing off Uruguay they were intercepted by HMS *Exeter* and told 'to get out of the way' as the Royal Navy was attempting to fight the German pocket battleship *Graf Spee*.

Rick and his Polish ship headed for Southampton, having called in at Dakar in West Africa. On joining the Free Polish Forces in France, he was put into a cavalry regiment until the authorities decided he would be far better employed as a sailor. Ryk therefore joined the Polish Navy on board the *Kosciusko* at Devonport, from where he sailed to Malta to join HMS *Garland*, a warship with a mixed Polish and British crew. This was in the spring of 1940 and *Garland* sailed to and from Russia until May 1942 when, in a convoy to Murmansk, a Junkers JU 88 dive-bombed the ship and scored a direct hit. The *Garland* was crippled and Ryk received severe head wounds. He was picked up out of the water, taken to hospital in Murmansk, and thence to Archangel from where the United States Navy evacuated him to Glasgow. He was discharged from hospital on 26 June 1943 and awarded a 90 per cent War Pension immediately.

Rick went to Birmingham and remained for eighteen months in lodgings paid for out of his War Pension, unable to work at all. Eventually he went to the BSA (Birmingham Small Arms) Company where he learned to become a precision grinder and thence to the Rover Car Company where he worked for twenty-seven years until he retired. During these years his War Pension was gradually reduced to 30 per cent. After the war he married the lady with whom he had lodged since leaving the Navy.

After retirement Rick began to suffer from the headaches, deafness, depression and anxiety – problems he had suffered from for most of his life. When I met him, he was worrying about his health and suffering from more than his fair share of bad days. Ryk has benefitted from regular stays at Tyrwhitt House, and he looks forward to further stays. His War Pension has been restored and the Society has found Rick to be something of an inspiration to many of the other clients.

Lee volunteered for the Army in September 1939 and found himself in the First Maritime Regiment, Royal Artillery. They manned the guns mounted on the decks of merchant ships sailing in convoys or in support of our Land Forces. Lee proudly showed me his medals; he held the rare combination of Atlantic Star,

North Africa Star and the Burma Star. He had served in a capacity similar to that of the Royal Marines except that he was on merchant vessels. After convoying across the Atlantic, where he was torpedoed, Lee took part in various landings off North Africa for two years before being sent to support landings off Burma for another two years. Lee had risen to the rank of Sergeant.

At this point, Lee suffered a nervous breakdown on being told that he was to be transferred to the convoys which went round the North Cape to Russia. He described to me how he told his commanding officer that he had had enough and tore off his sergeant's stripes from his tunic. This would have been a court martial offence but for the fact that the CO was an understanding man. He realised that Lee had a first-class record of bravery under fire off Egypt, Greece, Ceylon and Burma up to 1944. Lee was reduced to the ranks and granted a medical discharge with exemplary conduct. He was granted a 20 per cent War Pension which had never been changed when I met him in 1992.

Lee, on leaving the Army, worked quietly in South Africa for the next thirty years before he came home and became a part-time doorman at a seaside theatre on the South Coast. Lee was made to retire when the management realised one day that he was eighty one! He now lives alone in an upstairs flat, never takes a holiday and knows he is forgetful. Lee never received a proper Old Age Pension because of his years working abroad. He did not qualify for a South African pension either. Fortunately we were able to encourage him to apply to have his War Pension reviewed and, when I last saw him, it stood at 40 per cent and he was benefitting from a little extra money. Lee declined to take a convalescent holiday, saying that the physically disabled were more deserving than he was.

Bill Matthews and his wife became more like friends than clients. I met them in 1986. Bill had had to give up work in 1971, in his early fifties. He and his wife lived in a large house where the heating bills were expensive because Bill was at home for most of the time. They had a small War Pension. He was cheerful and friendly but obviously a bundle of nerves and very dependent on his wife, who cared deeply for him.

Bill was a pre-war regular sailor who served first in Palestine in 1938. He saw serious action off Narvik in 1940 when his ship was

Petty Officer
William Matthews RN.

Bill and Becky
in recent years.

badly damaged. He then took part in the Russian convoys until he was sunk. Later Bill took part in mine-sweeping raids on Dieppe and Saint Nazaire, all of which were extremely stressful. However, Bill survived it all and rose to the rank of Petty Officer. But worse was to come, in 1944. He survived the Normandy landings and by Christmas 1944 they were in Antwerp, in dry dock from which they could not escape by sailing away, when the port was attacked by flying bombs which caused massive fires and explosions all around. Many people were killed. Years later, he could still recall it all very clearly. He went on leave and, for the first time ever, he determined that he did not want to go back. His personality had changed, and he became frightened by everything and everyone.

As a regular, Bill stayed in the Navy after the war. Informally, he had been advised to leave, in the expectation that he would improve once he was out of the Navy, but he was not invalided out. Nearly half the crew of Bill's mine-sweeper were found to have tuberculosis. He was posted to Gibraltar after the war, as Petty Officer in charge of naval stores. One supposes the Admiralty thought this would be a quiet posting for someone who had endured so much on active Service. This was not to be the case and Bill's mental state turned him into a perfectionist and he became convinced that everyone else was stealing from the Navy, whether this was true or not. Bill's former anxieties had returned and he was put on tranquilisers. When he was due for his next naval medical examination, in 1949, Bill was recommended for discharge.

Bill's civilian career suffered because of his nervous anxiety, but he carried on as a storekeeper for the Electricity Board until computers made him redundant. A similar job at an engineering firm ended when Bill began to have blackouts and circulation problems.

Bill Matthews died only recently, but he was happy to benefit from the Society's help. His pension was changed from being for an aggravated condition to being attributable to the Service and was increased accordingly. The arrears of pay went towards insulating and heating the house. Bill enjoyed several periods at Tyrwhitt House and perhaps the highlight of the help the Society was able to arrange was a holiday for the two of them, in Bournemouth, paid for by The Not Forgotten Association. After Bill's death, the Society was able to help show that his death was associated with his war pensionable condition. Consequently, Becky, his widow, was entitled to receive a portion of his war pension for the remainder of her life, until she died in 1998.

Stoker **Bill Brown** sailed all over the world on destroyers and frigates. He was always below decks, working in appalling heat, especially in the tropics. In such conditions tuberculosis was not uncommon, then there were the horrors of being stuck in the engine room if the ships were damaged or sunk through enemy action. Having to work and live in a confined space was to take its toll on the stoker.

After the war Bill could not settle to any job which required him to work indoors so, after several changes of occupation, Stoker Bill became a milkman for a number of years. Eventually nervous anxiety overtook him to such a degree that he could not concentrate sufficiently to deliver the right milk or give the right change to his customers. When Bill lost his job he became depressed; the more he stayed indoors the more he became agoraphobic until he acquired a second-hand mini car. Going out in the car became his means of coping with his condition and, most days, he would drive out into the country, a few miles from home, breathe in the fresh air and stretch his legs.

When I first met Stoker Brown he was becoming depressed again. The car needed to take its MOT test and he expected bills he could not pay. He was a worried man. Enquiries showed that he did not qualify for assistance under the War Pensioners' Mobility Scheme because, officially, Bill was mobile physically. We were able, however, to obtain small grants from the Royal British Legion (RBL) and the Royal Naval Benevolent Trust (RNBT), and the car was repaired. I was able to repeat the grant exercise, although Bill did not like taking from charities. These charities do not generally give regular grants to a client, they much prefer to fulfil one specific need at any one time. After all, to have undertaken to keep Bill's car in repair would have been like giving him a blank cheque.

In the meantime, Stoker Brown went to Tyrwhitt House where he was recommended for increases to his War Pension. With a little more cash in his pocket, he became more cheerful. Recourse to the charities for help in running his car were now not so frequent or necessary.

John was a porter on the Southern Railway before the war. He joined the Navy in 1940. After basic training as a seaman, John was on continuous active Service at sea until medically discharged in 1943. On leaving the Service he was suffering from nervous anxiety and awarded a War Pension for an aggravated condition. John had been mentioned in dispatches for his bravery on HMS *Brocklesby* in the raid on Saint Nazaire in May 1942. *Brocklesby* was an old destroyer, manned by a skeleton crew, packed with explosives and rammed into the dock gate at Saint Nazaire with the purpose of blowing her up and putting the port out of action. The crew scrambled ashore, and planted more explosives in the harbour, before being picked up later by other ships of the Royal Navy.

By 1944, John was in a psychiatric hospital, certified as insane, and being given electro-convulsive therapy (ECT). He was a patient there for the next thirty years, eventually working in a factory outside the hospital. John was made redundant when the factory closed.

I found John living in a hostel for the disabled from where he was moved to supervised lodgings under the 'Care in the Community' policy. There, John found himself trying to cook for others in a worse physical condition than he was himself. Then there were the long weekends when there was no supervision at all. His sister became concerned to find John and his companions living on fish and chips on Saturdays and sandwiches on Sundays. John's condition deteriorated, and it was discovered that he was not taking his medication.

I found John polite, friendly but slightly confused mentally. The Society was able to place John in a private psychiatric nursing home on the South Coast where he stayed happily until he died only a few years ago.

Lewis went to sea with the Merchant Navy as a boy of sixteen. He trained as a radio officer at the Wireless College, Holloway, before the war. He served on North Atlantic convoys, voyages to South America, and on supply runs to North Africa. Lewis was invalided out in 1943, suffering from nervous shock; he was only twenty-four. A typical situation was where he held the fort alone in the ship's wireless room while the rest of the crew manned the guns. Lewis was officially declared to be unfit to go to sea again and given a 70 per cent War Pension for psychosis attributable to Service conditions.

Lewis then undertook various jobs, including working for the BBC's morse monitoring unit at Caversham and the Admiralty Shore Wireless Service. His last work was with the Thames Conservancy Board as a lock keeper, but even this was too much and he had to retire completely at the age of forty-eight.

When I saw Lewis twenty years after this, he was living on his own in Devon. His wife had left him. Although he needed support and friendship, he was reasonably provided for financially. The Society arranged for him to drive himself to Tyrwhitt House where he could use his car for the benefit of others, besides helping in the occupational therapy department.

The saddest naval case I ever came across was that of an elderly man, living out his declining years in a private nursing home on the South Coast. **Ben** was reported to me by Matron, who found out that he had served right through the Second World War in the Royal Navy. He suffered from severe loss of memory but his eyes lit up when I showed him pictures of warships.

I was called in because Ben did not have any pocket money. The staff gave him some of their own sweets and cigarettes, but there was no money even for personal toilet items. His family had placed Ben in the home at a time when the Social Security benefits covered the fees. But eventually, the fees outstripped the allowances and, when invited to contribute, the relatives disappeared from the scene and stopped visiting. The proprietor then took Ben's pocket money, which he was allowed under DSS Rules, to go towards the shortfall in fees. He was reluctant to move Ben because it would disorientate him, although he was of course losing money by keeping him on. I tried to contact the family, but they never responded. A friend of mine in the Royal Naval Association arranged a visit, and a gift of a few pounds was made. The Society could not even commence proceedings towards a War Pension, or a grant from the Royal Naval Benevolent Trust, because his relatives held all the details of Ben's Service record.

I visited Ben many times, for as many years as I was able. I gave him a few pounds towards his needs but these circumstances filled me with sadness and frustration.

*   *   *

So far, I have only briefly touched on the War in the Air, with reference to the Battle of Britain. In the First World War it was found that RAF pilots had a length of flying efficiency which varied from man to man but which, if exceeded, could lead to a breakdown. Observers and gunners were more likely than pilots to be affected by nervousness whilst those who went up in aerial balloons, spotting targets for the Artillery, were even more likely to suffer from a nervous disability. I suspect there were feelings of helplessness while someone else actually flew the plane. Moreover, aircrew, other than the pilot, might have long periods of inactivity waiting for something to happen. When it did it was very sudden. The pilot was less likely to suffer from nervous anxiety because he was occupied during the whole of the flight.

He at least could see the danger coming up, and was, to some extent, prepared. For other aircrew, especially those without a view to the outside, anxiety levels could be extreme.

In the Second World War, the Germans sent their pilots to rest areas; the RAF sent pilots on leave if they had previously done well for long periods. Bombing operations were more difficult to control. Bombs could be jettisoned at sea, the plane's own guns could damage its wing tips and the log of the flight could be altered.

If such irregularities were found out, an accusation of LMF (Low Moral Fibre) could lead to the loss of the coveted wings badge. Some pilots lost their wings on a formal parade; in other cases, it would be done quietly. Aircrew personnel thought this system unfair because it was seen to be administered by officers who had not seen any operational flying. Many of those who survived were to develop PTSD in later years.

Under the American system, an airman who refused to fly would be confronted by the Medical Officer, who would attempt to persuade him to return to operations. If this did not work, then there would be a direct military order to fly and, if not followed, there would be disciplinary action. It was recognised that the sight of enemy aircraft being shot down could cause distress, and this factor was a part of the survivors' guilt syndrome.

*   *   *

Vernon could not wait to join the RAF. He left school in 1937 and became an airframe apprentice at fifteen and a half. Later he worked as a flight engineer on Lancaster bombers, operating from Lincolnshire. Vernon had completed thirty-one bombing raids over Germany when his plane crashed in East Anglia, on the way back home from a raid on Bochum. An engine failed as they flew over the North Sea, which Vernon could not rectify; then the altimeter failed, and the pilot thought they were flying higher than they actually were, so they crashed without warning. Vernon was unconscious in hospital for several days. He never flew again.

Vernon spent a period in an RAF psychiatric hospital before medical discharge in 1944. In 1945 he was a patient at Stour House near Dedham in Essex, a convalescent home established by the Ex-Services Mental Welfare Society in 1941. Stour House had been set up by the Society after the destruction of Eden Lodge Beckenham, in an air raid. It continued to take in patients until 1949, when Tyrwhitt House was established.

Vernon did a variety of jobs: he was a bus conductor, an airframe fitter and a process operator for Shell on the Isle of Grain Refinery. His final job was as an insurance salesman. He struggled along until 1984, suffering the mood swings he had encountered since 1966, with periods in mental hospitals. Subsequently Vernon became a manic depressive with sudden changes in mood from optimism to severe pessimism, none of which he could control. When I met Vernon in 1990, he was a hospital out-patient. He slept badly, his memory was patchy, he was bad tempered and always apologising to his long-suffering wife who had herself become affected by it all. A bad mood might last for several months. We took Vernon into Tyrwhitt House, but he was unable to derive any benefit and in his periods of feeling unsettled, he became irrational. Neighbours were concerned for both Vernon and his wife and he was in and out of hospital until he died recently.

One day I received a request to call urgently on a war pensioner who had been awarded a lump sum which he was proposing to reject as being insufficient. I gave him the advice I gave to anyone in this situation, that he should accept the money offered, then, after a further period of time, ask for a review, indicating reasons why the percentage ought to be greater than what had been granted. Further information regarding a possible deterioration in his condition together with further evidence, if available, relating to the original case should then be put forward.

The pensioner, **Derek**, accepted my advice and told me his story. He was a Warrant Officer Navigator on a Halifax Bomber when he was shot down over Germany in 1943. He was able to escape from the stricken aircraft by parachute before the doors jammed. Only two of the aircrew survived. After landing, he escaped capture and made for Holland. He travelled back to England via Belgium and Switzerland. Derek never flew again because he always felt guilty when he thought of the other five crew members who had died. Derek told me in detail how the Halifax caught fire and the pilot struggled to keep the plane level for as long as he could to enable Derek and his comrade to escape, but the brave pilot was himself killed in the crash. That pilot died to enable Derek to live and he can never forget him.

Derek took normal demobilisation in 1945 and went back to complete his apprenticeship as an engineer's wood pattern maker – a highly skilled trade. He eventually stayed with another firm for twenty-one years until he had to retire suffering from stress. When I met him, he was clearly in need of stress counselling. He was reluctant to consider Tyrwhitt House as this would entail leaving his wife. She was unwell and they relied on each other. After a while, his War Pensions Welfare Officer was asked to initiate a deterioration claim and his pension was increased to 20 per cent on the advice of his GP, after a medical board examination.

A problem arose when Derek was faced with trying to replace their old car which they used for shopping expeditions. The Society assisted in applying for a low-interest loan from the RAF Benevolent Fund, but we were informed that they could not help in spite of strong support from the GP. We were unfortunate in that all the ex-Service charities were having to exercise stringency regarding loans and grants because of increased calls on their limited resources. Many of their applicants are now retired and in poor health. Derek managed to get his car through the MOT test, yet again. And to the best of my knowledge, it is still running.

A War Pensions Officer referred me the case of Flight Lieutenant **Jeremy Spencer** who volunteered for the RAF in 1940 to become rear gunner/wireless operator with the Middle East Air Force in the Western Desert. In 1944/45 he was with the Pathfinders in Bomber Command flying over Western Europe. He completed about four years of operational flying. Jeremy did not want to relate how he won the Distinguished Flying Cross twice.

On coming out of the Service, he joined the Metropolitan Police where he rose to the rank of Superintendant before being obliged to take early retirement. Some years later he was awarded a small War Pension for tinnitus in both ears with associated nervous symptoms. Later, the Society was instrumental in getting PTSD added to his pension.

When I met Jeremy he was badly in need of counselling. His GP knew nothing about his problems, and indeed how he felt. Jeremy had never told him. I first had to encourage Jeremy to see his doctor again. By then he was suffering from agoraphobia, refusing to leave the house, to take a holiday or even go as far as the local shops. In addition, he had severe bouts of depression, and there were physical signs of stress including nervous pains in various parts of his body. He was afraid of loud noises and, in conversation, became very emotional. Some pieces of music always made him cry; any news of an aeroplane accident distressed him tremendously. Jeremy was afraid of answering the telephone. In spite of all this, when he showed me his work, I immediately recognised that he was a brilliant artist and sculptor of great talent.

Jeremy was fortunate in having a caring former nurse as his second wife; and he had a good pension from the Police. In view of his condition, Tyrwhitt House was not feasible and indeed he would never consider spending a night away from home. The Society arranged to have him seen by Surgeon Captain Morgan O'Connell at Haslar Royal Naval Hospital on a one-day visit. Jeremy benefitted greatly from this consultation. The next stage was to have him seen by Wing Commander Gordon Turnbull, a retired RAF consultant psychiatrist, now practising in Kent. When I last saw Jeremy he was much improved thanks to the therapy from Dr Turnbull. War Pensions had put him in touch with their Occupational Adviser, who promotes arts and crafts for the disabled. This lady had plans for Jeremy to help with painting and drawing classes for other pensioners. I felt that he might do something similar at Tyrwhitt House, if he travelled there daily.

Gilbert is a 100 per cent disabled pensioner who served in the RAF. He had a nervous breakdown at the age of fifty and has been depressed and suicidal ever since, for the past twenty years. He was a commissioned rear gunner in a plane which crashed on take-off fully loaded with bombs and fuel. Gilbert was blown out of his turret as the aircraft became a fireball; all the other crew members died instantly. He was picked up from the tarmac with fractures to almost every bone in his body. His maximum War Pension is for head injuries with headaches and severe shock together with multiple bone fractures. His mobility now is poor, brought on by arthritis, and he has spondylitis in the neck so that he cannot look down. He therefore cannot read a book or drive.

His sole pleasure is to listen to records or literature on recorded tape.

On leaving the RAF Gilbert went to Malaya as a tea planter, but sadly his wife contracted a rare, permanently debilitating tropical disease. She is now a very sick woman. They came back to this country and he took a desk job with the same tea company until problems relating to his multiple injuries prevented him from working.

Gilbert has tried every possible treatment and medication, including acupuncture, to relieve the pain in his back and neck. However, the most effective treatment was to have his spine manipulated by a chiropractor. This is expensive, and is not available on the NHS where Gilbert lives. He has spent more than he can afford on chiropractic treatment. I suggested that he request help from the RAFBF, but he was clearly too proud to seek charity. When I retired, Gilbert's case was still unresolved with no source of funding for the periodic treatment he so desperately needs. Gilbert is a friendly, unfortunate gentleman who has endured tremendous suffering as a result of his RAF Service. He and his wife deserve better than this.

The RAF also had its share of casualties on the ground. **Richard** was an apprentice at a garage before the war. When he joined the RAF in 1940 he became a Corporal Airframe Fitter. He served all over the Central Mediterranean from Algeria in 1942 to Corsica in 1945. Richard was in the landings at Salerno, south of Naples, in 1943. He took normal discharge after the war.

Richard returned to the garage, where he became foreman until, in 1969, he found he could not concentrate. From then onwards, he was in and out of hospital. But each time the firm kept on taking him back as a loyal employee. This went on until Richard was sixty-four.

I met Richard in the council flat he shares with his devoted wife. He receives support from the local RAF Association. For the past ten years he has been treated at RAF Hospital Wroughton for PTSD, associated with his War Pension. Richard's problems stem from the stress of servicing combat aircraft. He has nightmares, which relate to rescuing the pilot from a blazing Hurricane Fighter at Tangmere in August 1940. The Society was able to help him.

Philip was a Flying Officer Navigator in Bomber Command.

On one mission over Germany his Wellington bomber was badly damaged; two of the four engines were out of action and the pilot was seriously wounded. Philip managed to fly the plane back on the two remaining engines in spite of the fact that the rest of the crew blacked out at one stage for lack of oxygen. For this, he was awarded the Distinguished Flying Cross. He had previously won the Distinguished Flying Medal on an earlier mission. At the end of the war, until 1947 Philip flew with Transport Command taking personnel to and from India and Singapore.

On leaving the RAF he joined the Civil Service and became an Executive Officer, taking early retirement at the age of sixty-three. After a happy year in which they went to visit their daughter in America, Philip became ill and his life-style rapidly slowed down. He could no longer drive and was given ECT at a local psychiatric hospital. He was put in a private nursing home to give his wife a break.

When I met Philip he was back at home but his mental condition was deteriorating fast. He had a nervous stammer, and he suffered nightmares which related to his RAF experiences. He lacked concentration, and could no longer read as he was unable to remember the words he had just seen. He sat and merely watched the pictures on television. He also worried a lot over trivial matters.

The next time I called, it was because Tyrwhitt House could not consider him for admittance in the absence of a report from the local Consultant Psychiatrist. I personally explained the importance of the psychiatrist's support if the Society was to offer a place to Philip. The doctor doubted whether much could be done for Philip who, in the meantime, had been readmitted to the private nursing home as his wife could not cope. Their money was rapidly disappearing to pay for his care. I called on Philip and he was clearly bored, surrounded by elderly ladies to whom he could not relate. By then he had an incontinence problem on top of his other difficulties.

When I next called, Philip was back at home having had three weeks at Tyrwhitt House. Both he and his wife spoke very highly of the home. The RAF Association representative had called and he was satisfied that Philip and his wife were receiving everything to which they were entitled from the DSS. However, Parkinson's disease had been diagnosed by the Society's Chief Consultant at Tyrwhitt House. Six months later, Philip was back in a nursing home.

When I called again, Philip's wife told me they were extremely hard up and that he was now in a less expensive home. I provided Philip with a claim form, in order to request a grant from the RAF Benevolent Fund towards their expenses. I also strongly advised her to apply for a War Disability Pension through the Royal British Legion, with whom she said she was already in touch.

Three months later I actually helped her complete the War Pension application, together with a request to the RAFBF asking for a lump sum to be borrowed on the security of their house. The next time I called, I was happy to report that they were entitled to some additional Income Support and I suggested she tried for help from the Civil Service Benevolent Fund. The RAFBF had offered an interest-free loan but the War Pension application had never been sent off! Bernard was now said to be suffering from cancer.

Philip died in 1991. His wife never applied for a War Pension and she went to live with their daughter in America. I sometimes wonder, what more could have been done to help in this sad, deserving case? After all, Philip did hold the DFC and the DFM, a rare achievement for acts of gallantry.

I wish to close this section on RAF cases from the Second World War with a mention of **Ray**, who was one of my first clients. Ray died almost ten years ago and he left on me a lasting impression of loyalty, bravery and selflessness, in the highest traditions of the Royal Air Force. Ray had been a Flight Sergeant Flight Engineer on a bomber which was shot down over the Mediterranean. He and others had drifted for days in a rubber dinghy before being rescued. He was found to be suffering from multiple injuries.

Ray was medically discharged but managed to work, in engineering, after the war until the injuries to his head caused his eyesight to fail. He and his wife then had to rely on his War Pension to supplement the Invalidity Benefit they received. Their problems increased when his left leg had to be amputated as a result of gangrene caused by gunshot wounds. When I met Ray, he had recently fallen over, trying to walk, and had seriously injured his hip, which had to be operated upon. Ray, by then, was almost blind. Further complications included diabetes, and he suffered a slight stroke. There was also the fear of gangrene in Ray's other leg. Yet, he was a man with a remarkable Christian faith. When I offered him a holiday at Tyrwhitt House, he said, 'If I had to be taken everywhere, I'd only get in the way. Give my place to somebody more deserving.' There was no answer to this.

Ray conversed at length about the Air Force. He was very knowledgeable about all types of aircraft and their engines, past and present. When I told him about my son who is an airline pilot, Ray knew all about the aeroplane he was then flying, even down to where the controls were.

Ray died on 5 May 1985 and he was buried on the day after the fortieth anniversary of VE Day. Before the end came, he was in bed and the family had gathered all around. His voice changed to that of a young man and he went through a complete pre-flight fuselage and engine check on a Wellington Bomber. "Port outer . . . Port inner . . . Starboard inner . . . Starboard outer." Each engine was tested up to a certain number of revolutions per minute. Younger members of the family had no idea what Ray was doing but Kathleen, his wife, knew very well what it all meant. Ray was leaving to join his greatest loves, his God and his Royal Air Force.

Ray had requested an oak coffin and his wife managed to find just the right one. The church was decorated with flags for the VE Day commemorations. The RAF Association and the RBL all turned up and his coffin was suitably draped with the Union Jack. The RAF provided a trumpeter to sound the Last Post and Reveille. Ray received the send-off he would have wished and which he so much deserved.

Kathleen, his wife, received her share of Ray's War Pension. She returned to me the 'talking books' for the blind which we had borrowed for Ray when he could no longer see to read for himself. Of all my hundreds of clients Ray is the one I shall never forget. I will always hold his memory in the deepest respect. He was the sort of person who, as he lived and died, made all my work worthwhile.

*   *   *

The Second World War extended into the homes of almost everyone in Britain, as most people experienced air raids which could take away houses and families without warning. As my work was primarily concerned with ex-Servicemen and women I was surprised, at first, to encounter those who had been the victims of air raids at home, which may or may not have been connected with their duties in the Forces.

Many soldiers were stationed on anti-aircraft batteries which were themselves targets for enemy raiders. **Mike** volunteered for the Army at nineteen, and became an anti-aircraft gunner based at Scapa Flow in the Orkneys. Towards the end of the war he was blown up by a bomb. The injury left his hands and arms badly deformed. At the Rehabilitation Centre Mike became a painter and decorator, but he had to give up after two years. Subsequently his 100 per cent War Pension was restored to him. Mike then became a Civil Service messenger until he was medically retired at the age of forty-nine. He has had a history of blackouts and mental depression brought on by physical problems. His wife suffers from agoraphobia. The Society was able to give Mike a break at Tyrwhitt House and he returned home much improved.

Mary Chespy joined the Civil Defence, known as ARP (Air Raid Precautions) at the beginning of the war. She described how she was on duty in 1943 when she was hit by shrapnel in the legs, with severe burns and bullet wounds to her arm. Having recovered sufficiently, Mary appeared before an Injuries Compensation Board and was encouraged to join the Forces and to try to forget about her injuries. This was a surprising recommendation in the circumstances, but Mary did follow the advice. She passed her medical, was graded A1, and joined the WAAF where she became a radar operator with RAF Fighter Command. She was eventually medically downgraded and discharged as having a psychopathic personality (due to an incident where she once lost her temper). She worked for many years as a bus conductress and, after an injury at work, she applied for a War Disability Pension and received 20 per cent. She was never able to have children although she married in 1953.

Mary's War Pension was later amended to include fracture of arm, leg and psychoneurosis. She suffers from arthritis in her injured arm and has become diabetic; she struggles to overcome her fear of going out of doors. Her main hobbies are a love of operatic music and her amateur radio transmitter, which has given her

Mary Chespy at Tyrwhitt House and in her RAF uniform during the war.

friends all over the world. Mary has been a cheerful client who benefits regularly from Tyrwhitt House, and the Society was able to get her War Pension increased a few years ago.

Sam lived in South London and joined the ARP as a cyclist messenger at the outbreak of war, aged fifteen. He was in the basement of a building which was hit by a high explosive parachute mine. Water was rising as a result of fractured sewers and water mains and he was trapped for nearly seven hours. During this time, however, he helped to rescue several small children through a hole in the masonry.

When he was seventeen, Sam joined the Home Guard, whilst working in an aircraft factory. He joined the Fleet Air Arm as an Air Fitter in 1943 but was medically discharged one year later. He was awarded a 20 per cent War Pension for rheumatic fever aggravated by Service conditions. He was convinced his experiences in the Blitz had been the root cause of his poor health.

Sam went into automotive electrical work and did various other jobs until he had a major heart attack at fifty. As a result, he would not return to work for over a year. He has had three heart attacks and two heart failures in the last ten years. Sam is now diabetic, suffers from thrombosis and is, naturally, depressed. The Society took steps to have Sam's War Pension reviewed, as I was convinced that his report of what happened during the London Blitz may well be considered as a possible origin of his present mental and physical state.

One of the most dangerous tasks in the Armed Forces is that of working with Bomb Disposal Engineers. This work continues to the present day, not only in the event of a terrorist attack but whenever someone discovers an unexploded bomb, perhaps on a building site. Every assignment is fraught with anxiety and many personnel from the Royal Engineers who undertake this work have sooner or later been the victims of PTSD.

Norman was an apprentice bricklayer who volunteered for the Royal Engineers in 1940, in the hope of continuing in the building trade. The Army, however, had other ideas and Norman was trained as a Bomb Disposal Engineer, stationed in Yorkshire. He was involved in bomb clearance in Sheffield, Hull and other places. Pressure of work and its dangers took its toll. Norman suffered from loss of memory and began to shake with anxiety. He was eventually discharged from Northfield Military Psychiatric Hospital Birmingham in 1943.

Norman settled down at his old trade of bricklayer: in fact he did so well that his War Pension was cancelled when it was found that he was earning as much as he would have done had he never joined the Army at all! He saved up and he and his wife retired early to Devon. But his eyesight deteriorated and his pension was restored. When his wife became ill, Norman's confidence left him and his wartime fears reappeared. The Society has helped to increase his pension and to give him an occasional break.

My last client in this part of the book was in the WAAF (Women's Auxiliary Air Force). She had worked as a photographer before the war and was a Clerk at the RAF Reception Camp at Bournemouth. **Joan** was taking her lunch break on the beach when a German plane flew in low overhead, raking everything around with machine-gun fire. The girls dived for cover and when they arrived back at the camp they found serious damage to both buildings and personnel. The shock of this experience filled Joan with nervous anxiety, which led to a duodenal ulcer and to her eventual medical discharge. No War Pension was offered and Joan never thought to apply, even when she began to have fits and went to hospital for psychiatric treatment. Joan went back into photography and also began to deal in antiques. In 1950 she was in a psychiatric hospital for a year. When I met her, the business was not doing well and there were debts.

Joan's health worsened but she benefitted from the occasional stay at Tyrwhitt House. Eventually we obtained the necessary application forms and I helped her complete her case for a War Pension. We were successful even though the traumatic events we recalled took place some fifty years ago. The business was wound up, Joan's financial position is far from sound, but she is now receiving the help and support she deserves.

*   *   *

The award of War Pensions was not the sole aim of my work. The most important aspect of my work was to offer and build upon the friendships I made, thus encouraging the client to tackle his or her problems.

Clients were sent to Tyrwhitt House because it is a rare institution catering solely for mentally disabled ex-Servicemen and women. No other organisation is set up to cater for such individuals. The Society also has a similar home, Hollybush House, in Ayrshire, which is equally successful with similar patients. There is now a third home, Audley Court, in Shropshire, opened in 1996. Many other institutions, understandably, draw the line at accepting the mentally ill.

The Society often made gifts of clothing and footwear, some of which was generously donated by well-wishers including some of the clients themselves. The Society also arranges for grants – often from the appropriate ex-Service charity or a local charity – to be passed on. Sometimes approaches are made to a Local Authority, or landlord, for better conditions, a fair rent, or perhaps a move to somewhere more suitable. Then there are holidays and convalescent breaks for families. A GP or a hospital might be approached on behalf of a client unable to express his or her requirements.

So far I have told the story of those I was able to assist but there were also some who could not be helped. Perhaps they feared that

I had come to have them put away. Some ex-Servicemen felt, rightly or wrongly, let down by the NHS. Most were very grateful for the help the Society gave them. It was also my job to put clients in touch with other agencies who were locally based, and better able to give direct help than I was. For example, other charities had funds available to which the Society did not have access.

Many times, old soldiers of both sexes still retained pride and affection for the regiment or Arm of the Service. Many felt they had never left, in spirit. Some did not want the regiment to know of their plight, and this could make it difficult to arrange the necessary help.

Old soldiers do not always get on with much younger nurses, doctors or social workers, although a military psychiatrist would always succeed in making some progress with them. The War Pensioners' Welfare Service eventually ran out of officers with Service in the Armed Forces and, in my time, I saw the arrival of welfare officers who were much younger than the clients from the Second World War. Many of the younger officers were ladies and they have all been absolutely first class in dealing with my now elderly veterans.

Periodic revivals of interest in the Second World War in films and television continue to bring it all back to those who sometimes prefer not to remember. In my experience, PTSD has definitely been sparked off by coverage in the media. Ex-Servicemen might discuss with me the inaccuracies of a film but, underneath, they might have relived it all over again, to the distress of those around them who might have very little idea abut what had actually taken place.

Many of my clients kept going to the end of the war, even though they felt very ill at the time they felt it their duty to do so. The regular soldiers continued for some years afterwards. Many worked for years before they were overtaken by permanent illness; their relatives often had no idea of the cause of it all. Thousands of such cases still exist.

We have seen how some of the ex-Service charities came into being, and how they were financed. To some, the arrival of the welfare state, able to care for everyone from cradle to grave, made all the other caring agencies redundant. In fact, the exact opposite was to be the case, as shown by events over the next fifty years following the Second World War. So great are the problems of the ex-Serviceman that no single agency, private or public Service, can possibly cope with them all.

Notes

[1] F. Spencer Chapman, *The Jungle is Neutral*, p. 209.
[2] *The Price of Surrender*, chap. 34 – 'Food for Seven Days', pp. 256–63.
[3] Geoffrey D. Vaughan. *The Way it Really Was* (The Granary Press, Budleigh Salterton, 1985).
[4] Terry's story has some similarities with that of Spike Milligan (see chapter 10).
[5] These centres will be discussed in chapter 5.
[6] R. H. Ahrenfelt, *Psychiatry in the British Army in the Second World War* (Routledge and Kegan Paul, 1958).

5

Demobilisation, Resettlement and the Military Hospitals

In spite of all the advances in knowledge regarding the discovery and treatment of tens of thousands of Service personnel with mental illnesses, the *Handbook of Army Health 1950* is a surprising official document by what it omits to mention. The Introduction states that 'the main causes of discharge from disease, in order of frequency, were (in the Second World War) psychiatric disorders, ulcer of the stomach, tuberculosis, bronchitis.[1]

The book further explains that from September 1939 to August 1945 more than 450,000 military other ranks died or were discharged from the Army on account of disease (including mental illness), compared with about 100,000 killed in action and less than 40,000 discharged as a result of wounds. These figures did not include those admitted to hospital and returned to duty, some with a lower medical category.

The Handbook, issued to all military units and medical personnel, went on to cover at length how the Army could combat this loss of manpower through sickness; the aspects covered were accommodation, nutrition, clothing, working conditions, recreational facilities, personal hygiene and welfare. Special attention was paid to water supplies, the control of pests, and health in the tropics. There were to be more routine inspections of barracks, the promotion of healthy exercise and the creation of job satisfaction. There followed an explanation of the Army's recent system for the medical classification of its men by the 'PULHEEMS' method.

P   = Physical capacity or general health
U   = Upperlimbs

L   =   Locomotion or lower limbs
H   =   Hearing
E.E   =   Right and left eyesight
M   =   Mental capacity for work determined by a series of intelligence and aptitude tests
S   =   Stability or the ability to withstand mental stress

Little was said in the 138-page book about the 'S' factor except under 'Man management' towards the end of the work.

Man management includes a short section on 'Mental Health' and this mentions five basic elements required for complete mental well-being. These are given as; *First*: physical needs, food, water, shelter and sexual contentment. *Second*: security or freedom from worries about work, postings, family matters, and being informed about what is happening in the unit and in the world outside. *Third*: social approval, which means that one's comrades and one's superiors make the individual feel a wanted and valued member of society, even though his job is a humble one. *Fourth*: competition and the opportunities it brings to show one's experience and skill, whether through sport or other factors such as inter-unit rivalry. (Fairness is deemed very important.) *Fifth*, and finally: mental health thrives on some degree of creative activity; some achieve this through their work while others find outlets in games or hobbies.

Thus the War Office was putting forward a positive prophylactic attitude towards mental health. Certainly the War Office was aware of the thousands of mentally sick soldiers from 1939 to 1945, but the philosophy of care was looking forwards, not backwards. Of course, nothing was said that might harm recruitment. Millions of men served two years National Service in the Forces. And at a time of near full employment, very few volunteered to stay on. There was, in my opinion, always a shortage of regulars even though the Army gave them better pay than the national Servicemen. The country was tired of war and in the post-war boom there were plenty of jobs to go to.

During the Second World War and just after, the Services made tremendous efforts in the treatment of mental illness. It was recognised that there were thousands of sick personnel who were casualties of battle, or of prisoner of war camps, or who had witnessed atrocities. But suddenly it was all abandoned, along with the demobilisation centres from where the men had collected their civilian suits, hats, raincoats and ration books.

In 1939 it had been decided that the treatment of neuroses was to be preferred to handing out pensions and simply discharging the patients. Manpower was to be conserved at all costs and, importantly, it was not to be made easy to avoid conscription by becoming a psychiatric case. Emergency Medical Service hospitals were set up by taking over parts of some civilian hospitals for Service casualties. Civilian consultants who were there were given honorary commissions in the Royal Army Medical Corps to preserve the military disciplinary hierarchy. Quite clearly, the old 'D' Block at Netley was going to be quite inadequate for all the military cases of psychoneuroses expected during the war. Some EMS hospitals had psychiatric out-patient centres and the military system had one great advantage over the civilian: patients under Military Law could be retained in a mental hospital without the need for certification, often for their own good.

Banstead Mental Hospital in Surrey had a military wing, set up in March 1940 for both Army and Air Force patients. In May 1941 a new scheme was devised for dealing with psychiatric patients discharged from the EMS at Mill Hill, North London. This was the 'annexure scheme' by which men were returned to an active unit, but for specific duties only, commensurate with their mental condition.

In April 1942 the largest Special Army Hospital opened at Northfield, Birmingham, with 200 treatment beds and 600 beds in a training wing for the rehabilitation of psychoneurosis cases for RTU (Return to unit). Another significant development was the establishment of the Directorate of Army Psychiatry in April 1942. Besides the usual department for the selection and appointment of military psychiatrists there was a branch which liaised with the Ministry of Labour and National Service, which ran the system of conscription into the Forces; there was also a branch which dealt with the Ministry of Pensions, handling discharges and medical boards. Thus it was recognised that the selection and discharge of military personnel did have an important bearing on mental health, at a time when every available man was required for the war effort. In January 1942 psychiatrists were employed at War Office selection boards where commissions were awarded. In the Second World War psychiatric problems were the largest cause of medical discharge from the Forces. Eight per cent of recruits were found to be dull or backward. Most of them went into the Pioneer Corps, where they would do labouring jobs.

The reaction of the Service chiefs to the employment of psychiatrists was, at first, not favourable. Churchill was sceptical and said, in December 1942, 'There are quite enough hangers on and camp followers already!' General medical opinion was also unfavourable, but the regimental medical officers who had to deal with day-to-day problems at unit level saw the point of it all. Some critics felt that the Army was losing too many men through discharges or special postings for men trying to evade active Service. This was the old problem from the First World War: how to separate those genuinely sick and incapable from the cowards and malingerers. The Southborough Committee had said in 1922 that the Services should only take on men with at least average mental health and stability. However, the Army had recruiting problems, caused by large numbers of skilled men at home in reserved occupations vital for the war effort, including manufacturing weapons and munitions. The Navy and the RAF tended to take the best recruits, as did Civil Defence. So it was that a large proportion of the country's 10 per cent of the population, who had mental or learning problems, found themselves in the Army. Some of them had been satisfactory doing labouring jobs at home when there was plenty of work to be had. But when they were faced with military discipline and stiff training, problems arose. On the other hand, some senior officers welcomed some of the dull men who were docile – at least they did as they were told! Some of the brighter ones attempted to obtain a psychiatric discharge in order to earn more money at home on shift work. As the war progressed, many commanding officers became more enlightened, realising that psychiatric cases simply did not mix well in a fighting unit. By 1943 men in medical category 'C' with psychiatric problems were discharged as unemployable; at the same time, Army selection centres were used to transfer men from the Navy and RAF to the Army. It was found that 18 per cent of the transferees were rejected as infantrymen. Thus the idea that the Navy and the Air Force took the best recruits was not always true during the war.

Those genuinely dull were a problem once it had been decided to retain men of low intelligence because of manpower shortages. They were misfits in their platoon; they became malingerers, deserters, a source of general discontent, as well as becoming physically ill, probably brought about by the fear of trying to cope with tasks beyond their capabilities. It was estimated that 4 per cent of recruits were totally unfit for Service and that the next 5 per cent were fit only for unskilled work in the Forces. (This was an estimate made in 1940.) Dullards formed a high proportion of psychiatric cases evacuated from France in the early days of the war. Hysteria could be brought on by basic military training; some of these individuals might have coped with the training of the 1914–18 war, but found the operation of the Bren Gun (Light machine-gun developed in the 1930s) quite beyond them, and thus extremely dangerous and hazardous for them to use.

*   *   *

I first met **Alec** in the West Midlands. He had five older brothers and sisters, all perfectly normal. Alec was looked after by his mother until she died. He was backward, living in semi-independence in a council flat on his own, with his older brother and sister-in-law, in their seventies, keeping an eye on him.

On leaving school Alec worked at the Austin factory in Birmingham but all he ever did was cleaning cars before they left the works. The Army took him assuming that he would be suitable for training as a vehicle mechanic, but he was found to be quite incapable of any form of skilled occupation. Alec was re-allocated to the Pioneer Corps as a labourer, but again he was not a success. He was eventually discharged after a lengthy period at the Military Psychiatric Hospital, Netley. In all, Alec was a soldier for two years as the Authorities were clearly reluctant to let him go. There was never any case for a War Pension.

I found Alec to be harmless, basically sensible but with hardly any speech unless pressed into conversation. He took a newspaper every day but I had no idea how much of it he understood. He could just manage to write his name. He had no idea of the value of money so his brother ordered the shopping and paid for it for Alec to collect each week. He did simple cooking and enjoyed a weekly pint of beer in a pub with his brother. He told me he had funny pains in his head; he walked to the local shops where he was known, but he was afraid of stepping on a bus.

The Society arranged for him to come to Tyrwhitt House, where he spent his time mainly on his own. His brother was pleased to have Alec looked after for a fortnight. While he was away they decorated his flat. However, when I last saw him, Alec was quite sure he did not wish to return to Tyrwhitt House. Perhaps he was afraid of being kept there indefinitely. Certainly his brother was pleased to know that a break was possible and that, one day, even Kingswood Grange for elderly veterans, might be appropriate.

Alec never worked again since leaving the Army and naturally, his brother was anxious for his future as none of the rest of the family wished to be involved with him. Alec was hardly a war hero and it is doubtful whether the Army worsened his condition at all. This case raises the question of whether the ex-Service charities ought to devote their time to men like Alec. I found it impossible to draw a line between men injured in action and those disabled in other ways during their Service, or since discharge. I found that most of them could benefit from the atmosphere of an ex-Service institution, in company with men and women of their own age and with many shared experiences. I feel that, in time, Alec could become accustomed to further stays at Tyrwhitt House where he would certainly be made welcome.

*   *   *

In 1943, it was ruled that a military psychiatrist should advise every court martial for disciplinary cases. Subsequently it was found that 80 per cent of individuals up on a court martial were psychiatric cases, and as many as a quarter of them were mentally defective. The problem remained that the authorities were reluctant to release psychiatric cases even though they were a danger to their comrades and frequently absent. It was also feared that large numbers would commit crimes in order to get out or, if kept in, the psychiatrist would 'get them off' from being punished. The Service chiefs did not wish to see the Forces regarded as a penal colony or forced labour camp. However, the majority of psychopathic personality cases were not discharged until 1944. An Army Council Instruction of 1945 authorised the discharge of habitual bad characters and psychiatric delinquents, and the discipline of the Army improved as a result. Many of these cases had had long histories of maladjustment at home, school or work. No military penal companies were successful in rehabilitating large numbers of misfits. In spite of a punishment of three years' penal servitude for desertion during the war, 15 per cent of deserters were found to be totally unfit for military Service. A Review of Sentences Board in Germany in 1945 found that, out of 2,000 soldiers tried for desertion, 57 per cent were psychiatric cases, immature, dullards or anti-social psychopaths. Many were below average intelligence. A quarter had fought in previous campaigns and their endurance had been stretched too far. There was no difference between regular Servicemen and conscripts, as selection had not been very stringent before the war. Many had fought since 1939 and domestic stress, poor leadership and lack of regimental spirit were given as reasons for desertion. Three-quarters of all military offences were for absence without leave.

One development worth mentioning was in Italy in 1943, where deserters were sent to a rehabilitation centre which aimed to send offenders back to the front line rather than to prison. This succeeded because offenders were given three months in which to redeem themselves and sentences could be suspended. Once again, the Authorities decided that there should be no evasion of military Service through misconduct.

*   *   *

Jeremy was a client with psychiatric problems. He had served with one of the Welsh regiments in India in the 1930s, where he was Acting Company Sergeant Major. He was discharged in 1938 suffering from nervous anxiety, trying to hold down a responsible position, in an alien land far from home, caught in a triangle between the officers he served, the men under him, and the native servants, in a situation where he felt he could never relax his guard. The mental strain proved too much for him.

Jeremy missed the Army when he came out and when he reached home, war was imminent. The only way he felt he could rejoin the Service was to volunteer for a different branch of the

Army under a false name. When he again began to feel unwell he was transferred to the Pioneer Corps where he settled down and quickly rose to the rank of Sergeant. This appeared to be a sensible move until the signs of nervous breakdown reappeared yet again and he was medically discharged for the last time, in 1943, with a 20 per cent War Pension for psychoneurosis aggravated by Service conditions. Jeremy never dared appeal for a higher pension for fear of his earlier history being discovered. For the rest of his days he felt that if only he had been able to stay in, with all his Service to count since 1933, he could have retired comfortably with the full pension of a Warrant Officer.

Jeremy only ever did labouring jobs, although he looked every inch like a retired Sergeant Major. He frequently benefitted from Tyrwhitt House and the Royal Pioneer Corps Benevolent Fund were very kind to him.

* * *

Soon after the war broke out, the Army's psychiatric Services began to expand. It had already been decided not to discharge automatically everyone pleading mental illness as an excuse for not going on active Service. Rehabilitation within the Army was to be the aim of the Emergency Medical Services neurosis centres. At Mill Hill and Banstead staff were posted in from the Maudesley Hospital in Denmark Hill. Half of those treated went back to their units although EMS hospitals were not, strictly speaking, military hospitals. Northfield Military Psychiatric Hospital Birmingham opened in 1942: it had two basic functions. The Training Wing aimed to return men back to their units, and thus all the 'patients' wore khaki uniforms. The Hospital Wing patients wore hospital blue uniforms, with the familiar white shirt and bright red tie. In the Training Wing there was some modified military training and there was some treatment from the psychiatrists in the Hospital Wing on an out-patient basis; they were all initially regarded as being capable of RTU (return to active Service unit). Many who were transferred from training to the hospital wing were thought of as malingerers.

Group therapy began in the Hospital Wing in 1943 and by 1944 young soldiers straight from Normandy were being treated there. By November 1944 the celebrated 'Northfield Experiment' began in earnest. In brief, the two hospital wings combined and became a self-governing community with expanding facilities for occupational, social, recreational and educational training. It was only to be expected that there would be some problems when all of this was felt to impinge on the authority of the Commanding Officer, who happened to be a Regular Army officer. After VE Day the emphasis changed from military rehabilitation to civilian rehabilitation and occupational therapy was further developed. A few battle shock casualties had been successfully sent back to the front line but many others had gone to lighter duties in the Army. As the war came to an end, the large numbers who were successfully discharged into civilian occupations were seen as a cost-cutting exercise, whereby War Pensions could be avoided.

The community hospital concept at Northfield merits some further study. The idea of communal activity had developed from the intellectual socialism of the Fabians and the rise of the Trades Union Movement and the Labour Party. The General Election of 1945 showed just how far public opinion had gone in this direction when the country elected, for the first time ever, a Labour Government with a majority large enough to implement a programme of public ownership on a large scale. Northfield reflected this atmosphere and was very significant in cutting across existing medical practices in which the doctor had always known best and the patient knew nothing. It could also cut right across military practices where the soldier was expected to do as he was told without question. From Northfield there arose a complete methodology of psychiatric treatment, which came to be known as the Group Approach, as opposed to the more traditional doctor–patient relationship, which had always existed.

The Northfield Centre took its ideas from the Tavistock Clinic where W. R. Bion believed that a group of patients could be treated together as an entity. After all, Service patients were accustomed to being part of a team, but they could feel inhibited by military discipline and this may not aid their recovery.

A group of patients could be set an allotted task, such as discussing an issue and reaching conclusions, or planning and carrying out an actual piece of work. Individuals who did not benefit from such an exercise would qualify for individual therapy as a follow-up. Some successful groups, since Northfield, have been groups of out patients meeting regularly, and who do not otherwise know each other and therefore do not inhibit each other's behaviour within the group. It may be necessary to select members who will interact with each other. Group therapy can lead towards an awareness, by the patient, of his or her condition and behaviour; awareness can lead to control, and hope can be shared with the others. Patients feel they are not unique and can begin to derive satisfaction from helping each other. The leader of the group must work to a plan otherwise the sessions may become merely a ritual airing of grievances with no development of the patients' insight or attitude. Each individual needs a combination of involvement, control and affection.[2]

Any institution for long-stay patients is a microcosm of society. The old-fashioned mental hospitals of the nineteenth century were usually enormous buildings, in some isolated place, catering for large numbers of patients with varying degrees of disability. We now realise that many who were near normal should have been better cared for in the community, but they became institutionalised, permanently accustomed to a routine which never changed and became their life's pattern.

* * *

Arthur had been in a West Country mental hospital since 1946. I saw him forty years later; he had simply been 'forgotten'. Arthur was an infantryman who had served right through the Second World War before being sent back to UK from Normandy suffering from battle fatigue. Northfield tried to rehabilitate him but without success, and he ended up in a large mental hospital doing odd jobs.

When I met Arthur he said he was the Head Gardener. He received no visits from anyone and he was therefore very pleased to tell me all about his work – some of it real, some of it imaginary. In fact, he tended a small vegetable plot and his produce was sold to visitors in the hospital shop. He received a few pounds a week for his labours and insisted on buying me some small presents for my grandchildren. In return, I gave him a pair of new shoes from the stock I always carried in my car. Arthur was on little or no medication, required no supervision and was free to go outside the hospital, but he never did. He was a polite, friendly man and

I often wonder what became of him. He could have foundered if ever he was discharged outside and left to fend for himself. That would simply not have been either appropriate or fair. I just hope he was allowed to stay inside the hospital to live out his life with some dignity.

The large mental hospitals of the nineteenth century, we are told, treated their patients kindly, as indeed was Arthur. Two-thirds of patients were discharged and all were found work; some went into Service on the large estates of the nobility. Such opportunities no longer exist today. All the large hospitals had theatres; with a view to therapy. Patients took part in the productions which were put on for the benefit of the other patients. The 'actors', by participating in group performances, learned skills and developed responsibility and independence – all of which helped improve their mental state.

An enquiry into the running of a mental hospital in 1976 showed that conflicts between staff and patients, or the patients themselves, led to violence. This was often due to poor communication. An authoritarian regime with inter-staff rivalries was more likely to be a cause of friction than were inadequate facilities. It was not clear whether the hospital was aiming to cure or merely just control the patients. Some patients were crying out for attention and would even become incontinent in order to obtain it. I have encountered this myself, and such behaviour ceased once the client felt secure in that it was recognised and felt, that someone cared for him. Where the atmosphere on a ward became relaxed it was very important for senior staff to become involved otherwise junior staff became too familiar with patients or distanced themselves from them. This is a basic fact of life in nursing homes as well as in large hospitals. Patients need to take some responsibilities, but the staff need to show no favouritism to any of them; this is difficult to achieve but it can be done with skilful leadership. In the survey, it was found that two-thirds of all long-stay patients were schizophrenics suffering from social and occupational inadequacy. Quick therapy does not work here. Control and care are needed and the sheltered hostel can be appropriate.

The idea of the 'therapeutic community' was first broached in the 1940s. In the First World War, shell-shocked victims had been given pensions for life and were regarded as permanently useless creatures who were organically injured. The British military psychiatric hospitals succeeded basically because the patients had a lot in common and they were accustomed to military discipline. The patients were receptive to taking new ideas at face value, after years of doing what they were told without question.

In fact, Northfield became anything but a military unit when it began to be run by the community rather than by the doctors, who changed their role for the purpose of the experiment. It was described as the socialisation of some very selfish people towards a stable life in the real world and some of the principles, revolutionary in the Second World War, are still practised today.

The new system at the Mill Hill Military Neuroses Unit developed from lectures to patients into their psychosomatic symptoms, which further developed into open discussions; this, in turn, developed into an unstructured debate on some of the everyday living problems the patients were having to face. Led by Dr Maxwell Jones, Mill Hill was based on ideas from the Maudesley Hospital. Work at Northfield proceeded on similar lines, and the system developed gradually. The lectures led to working in specific groups, such as returned prisoners of war; these individuals had problems relating to their acceptance by society, and how to obtain and how to keep employment. These patients all had much to learn about human relationships and the fact that the world did not owe them a living.

Finally, there was counselling in pairs where the listener did not join in but only prompted the patient to relate the hardest parts of his or her story. This is the method currently used by the Samaritans. The temporary military hospitals closed after the war but the ideas behind their pioneering efforts remain today. Group therapy continues in many aspects of military training and personnel selection.

*   *   *

Prisoners of war became such a problem in themselves that Civilian Resettlement Units (CRUs) were set up in March 1945, until they closed their doors in June 1947. There were twenty units, each taking 250 persons at a time for an average stay of five weeks. There were workshops for practising civilian skills, and some training and aptitude testing was carried out. Attempts were made to re-socialise the men through group discussions. Fifteen per cent of the other ranks and 25 per cent of the officers had psychiatric interviews. When psychiatrists had been employed in officer selection it had been found that psychiatric breakdown amongst the officers was more frequent amongst those in the lower ranges of officer intelligence and amongst those whose innate intelligence was higher than their educational standard; perhaps these officers had not been up to the job, or did not exhibit a rounded personality. The senior officers generally resented the use of psychiatrists at officer selection boards, and this practice was discontinued in 1946.

At the CRUs small group work was found to be useful in talking out problems and there were many 'freedom versus discipline' situations to discuss. One group of anti-social clients was warned that the rest of the CRU would vote for their expulsion and in this way they saw the effect of their behaviour on the others. Expulsion by the Officer in Charge would have sabotaged the democratic atmosphere of the place, but a vote by the insiders themselves had a different hue. Ideas and principles from Northfield and Mill Hill were prevalent in many institutions at this time.

The transition to civilian life was not easy to accept by men who had been fed, clothed and told what to do for up to six years; they now had to face up to new responsibilities. A review showed that 24 per cent who attended CRUs showed signs of social maladjustment whereas 64 per cent of those who did not attend exhibited unsettled behaviour. Those who had attended had gained an insight into their problems, especially at home; thus the CRU was a therapeutic community, able to help release men from some of their tensions and anxieties.

Similar problems arose from ex-Servicemen generally who had not been prisoners of war, especially when they returned from long periods overseas. The popularity of ex-soldiers at home was bound to be short-lived. A sample of non-POW expatriates showed that a quarter of them had POW syndromes. Servicemen had faced death destruction, horror, loneliness and they felt they knew more about life than civilians. Some felt that a grateful nation owed them a living. Left-wing politics encouraged this view, and they were irritated by civilians in positions of authority. Advice to 'pull yourself together' only led to contempt for the adviser; what did he know about what the victim had endured? When family or friends treated them differently from others, it hurt their feelings. The ex-Servicemen did not always appreciate that it was they who had changed.

After the war, the pendulum swung back as the lessons of two world wars were largely, or conveniently, forgotten. This was shown up in *The Handbook of Army Health 1950*. There had been problems of maintaining morale in the face of adversity, and morale is the only sound basis for discipline. This is true in any walk of life. Secondly, there was the experience gained in the prevention of psychiatric casualties, and in the best utilisation of all manpower resources. Thus there were three areas in which success during the war should have been built upon after the war, but it was not. During the war, far too much time was wasted in trying to cope with misfits in the Forces when the authorities decided that no one should escape military Service unless certified as insane.

After 1949 it was purely arbitrary as to whether a man was discharged under an Army Council Instruction (ACI) of that year, which stated that he was simply medically unfit for military Service; or discharged under an ACI of 1946, which was intended for delinquents who were also psychopaths! There never was a clear definition of the term 'psychopath'.

Notes

[1] Handbook of Army Health 1950. War Office Code No. 5691. Issued to all medical staff and to all units.

[2] J. S. Whiteley and John Gordon, *Group Approaches in Psychiatry* (Routledge & Kegan Paul, 1979). The theory is summarised as 'Forming, Storming, Norming and Performing' being the four consecutive stages in group therapy. See p. 82. The final stage is where the patient achieves insight into his problems leading to an improvement in his or her life.

6

Other Conflicts: Korea, Cyprus, Malaya, Aden

The year 1945 was a watershed in world history. It marked the end of the greatest global war of all time, and the beginnings of the Cold War between the Soviet Union and the rest of the Western Allies. After almost six years of war, Britain was impoverished financially. But, after the General Election of 1945, thoughts turned towards creating a new society within the concept of the Welfare State. The war marked the beginning of the end of the British Empire and the highest priority was given to demobilising the millions of Servicemen and women who had been conscripted for the duration.

It was blindly assumed that a grateful nation would welcome them home – that they would return to their old jobs, and blissfully resume their family life where they had left off. Many were to be disillusioned, for a variety of reasons. For the physically injured the problems were obvious; for the psychologically impaired it was far less straightforward. Perhaps the greatest difficulty was to lie with those who could not, or would not, accept that there was anything wrong with them. A related problem was that the place of women in society had changed, as a result of women undertaking men's jobs during the war. Many women were simply no longer prepared to become dependent again solely upon their husbands' earnings. Moreover, small children, left at home early in the war, were now teenagers when their fathers returned. Finally, it was assumed that the National Health Service would resolve everyone's medical problems. It was significant that rationing continued into the next decade and, in fact, bread was rationed for the first time, after the war – an indication of world food shortages.

Conscription was not abolished in 1945 for the simple reason that, as the wartime soldiers went home, there were not enough regulars to undertake all the military tasks which resulted from the ending of official hostilities. During the next fifty years Britain's Armed Forces served with distinction all over the world. The greatest conflict, for which they were always prepared, that of Communism versus Capitalism, to be fought in Europe, never took place. However, what did occupy our Forces came as a result of local nationalism in various countries, in various guises. There was also counter-terrorism, international peace-keeping, and of course acts of aggression in the Falklands and Kuwait.

The second half of the twentieth century was to be quite unlike the first. At the beginning of the century, Britain had looked upon itself as totally independent from Europe. And, with its own Empire, she was unwilling to become involved in other people's conflicts. When the British Expeditionary Force was first created in 1908, as a result of the new concept of 'Entente Cordiale' with France, it was a completely new departure. The BEF fought with distinction in France, sustained enormous casualties, and then withdrew from Europe just as soon as it was practicable. After the Second World War it was simply not possible to withdraw from Europe once again. Great Britain could not afford to police a world-wide Empire. And as the Empire was dismantled, so European involvement grew.

There were five million men in the Forces in 1945 and, after the majority of them had been demobilised at the end of the war, National Service was extended in 1947. It was to run for a further fifteen years.

There were immediate problems following the collapse of Japan, and prior to the arrival of Dutch and French troops to retake their colonies. In the face of nationalistic disturbances, following encouragement given by the Japanese throughout their occupation, there was no option after the war but to deploy the overstretched British Army to keep order in the Dutch East Indies and French Indo-China. This was in addition to reoccupying Singapore, Hong Kong and Burma. The French and the Dutch did not return until 1946.

There were similar nationalistic hopes in India which had to be faced when there were riots in Calcutta and the Indian Army mutinied. Prior, there had been promises of political independence in Britain's dealings with the Congress Party and the Moslem League. The final withdrawal from India came in February 1948, and this loss meant the end of significant British influence in the Middle East.

*   *   *

I met **Timothy** in the Birmingham area. On leaving school he had worked in engineering until called up, aged eighteen, in 1945. He was immediately sent to India and thence to the Dutch East Indies, where he and his comrades faced civil unrest. From there he went to Taiping in Malaya before moving on, yet again, to Burma where there was further unrest against the British. In October 1947 Frank was attacked by rebels in Rangoon and hit on the head by a rifle; he was taken to hospital suffering from loss of memory. He was considered for medical discharge, but Timothy said he was better than he really was, in order to avoid further tests and to retain his good medical record. As a result, no War Pension was considered.

Timothy returned home to his mother and to his former employers, but left this job after two weeks. He tried working on the land and in various engineering jobs, but could not settle down. His mental condition was so bad that his mother applied for a War Disability Pension in 1950, and he was awarded 30 per cent. By 1954 he had obtained light work in a metallurgical laboratory and had married. When I met Timothy in 1986 he was still working at the same firm, doing shift work and somehow managing to keep going. However, there had been spells in a mental hospital, one following an overdose of tablets, when Timothy had felt suicidal. He was given electro-convulsive therapy (ECT) and this led to some improvement in his condition. Tests revealed a blind spot in one eye which was regarded as having been caused by his head injury. No additional pension was awarded for this. Timothy was still getting pains in his head. In addition, there were other physical problems relating to this stomach and he was recommended for a hernia operation. Naturally, Timothy was worried about all his medical problems, the physical ones seemed to be made worse by his mental state.

Fortunately his caring wife was very supportive. She told me that Timothy never had a day off work except to go to hospital.

*   *   *

The lessening of British influence East of Suez had an inevitable effect on Palestine, where the situation quickly became intolerable for the British troops stationed there. When Jewish riots were put down, terrorist tactics were employed. The Army was so stretched that troops had to be sent from Germany to aid the Civil Power. There were sudden incidents, curfews, ambushes, strategic points to be manned and rioting at any time. Jewish refugees from Europe tried to enter Palestine by sea and the shore line had to be patrolled by the Royal Navy. In September 1947 it was announced that the British Mandate would end on 15 May 1948, whereupon the Jewish extremists and the Palestine Arabs openly began to fight each other. In all, 388 British lives were lost in Palestine at this time.

My clients from Palestine were not wounded in battle, as such. One was injured by sniper fire; another was the victim of the nature of his work in a stressful atmosphere where the British Army appeared to be friends of no one. He felt that he was denied the comradeship which, in similar circumstances elsewhere, would have aided his mental survival.

*   *   *

I found **Ted** in a group home run by MIND where he shared a flat with another patient. His marriage had failed and his daughter never came to see him. Ted volunteered for the Army in 1947 and was injured by a bullet above the left eye in an incident when he was a security guard on a train attacked by rebels in Palestine. This led to three months in hospital until his sight recovered. Ted then served in the Suez Canal Zone as a field gunner until his discharge in 1951. He became a lorry driver until he was forced to retire at the age of fifty-one, suffering from headaches which had led to a nervous breakdown and several spells in hospital. When I saw Ted he was an out-patient. He never considered applying for a War Pension although his headaches may well have been caused by the head injury he received in Palestine. Ted needs the Society's support and encouragement to have this matter looked into.

Roy was a grammar school boy who became an apprentice draughtsman, which in those days was the key to a successful career in engineering management. His father was a Police Officer and a stern disciplinarian. Roy was called up and, as one of the smartest young soldiers, was chosen to march in his regiment's Victory Parade in 1945. In January 1946, the Regiment went to Palestine. Whilst travelling on a crowded, slow-moving troop train, with open-sided carriages, a native infiltrator cut Roy's rifle sling, made off with the weapon and disappeared into the desert. The train was stopped, and later Roy was charged with negligence. His punishment, a heavy fine, was automatic. Two things now followed: Roy's father expressed his disapproval and, secondly, encouraged by senior ranks who should have known better, his own comrades forsook him. This state of affairs, against a background of terrorism where the British were unpopular with all sides, was to lead to Roy's mental and physical illnesses.

In one incident, an illegal immigrant ship, containing Jewish refugees from Eastern Europe, had to be emptied of its human cargo, at bayonet point. Conditions on board were appalling, women and children screamed, one had a baby on the open deck – all this shocked and upset Roy emotionally.

At this stage, some help appeared to be forthcoming in the form of a transfer to the Royal Corps of Signals as a cypher operator, with the rank of Lance-Corporal. In order to do this work, the operators were sworn to secrecy and lived in isolation from the rest of the camp. This was to become a frightening, lonely life until Roy went downhill physically. He was in hospital with dysentery for much of the rest of his Service.

Roy arrived home after a voyage on a troopship during which his money and his paybook were stolen. His father was not amused and the rest of his life has consisted of unsuccessful attempts to earn a living. His marriage failed, and ever since Roy has lived alone. He suffered from depression lasting for months on end, with periods in hospital for physical ailments.

Roy has kept going through an interest in music, especially jazz, and by a very keen interest in crime prevention in his local community (thus following in his father's footsteps). Over recent years, he has come to appreciate annual visits to Tyrwhitt House where he has been able to relax in a caring atmosphere. Persistent attempts to win the case for a War Pension have failed. In the early days after he left the Army, it never occurred to anyone to seek compensation for a promising career which came to nothing. This is a case which underlines the importance of comradeship in building and sustaining regimental morale and which was sadly lacking in Roy's case. In recent years there have also been problems regarding covering the cost of Roy's visits to the Society's convalescent home. This will be discussed, in more general terms, in chapter 7. This again has not helped Roy's morale. His problems were compounded by rejection at home because he had not returned as a war hero. Depression set in, which was to eventually ruin a promising life. Roy can still be deceptively cheerful which, in turn, probably did not help his case for a pension.

* * *

In Europe, British troops served in BAOR (Germany) Austria and Northern Italy, filling a vacuum of anarchy which would have quickly been occupied by the Russians who were refusing to accept the emergence from the war of a united Germany. Berlin was divided between the four occupying powers: Britain, France, Russia and the United States. The Western Powers supported their sectors of Berlin from the west by rail, canal and motorway. Pressure was brought to bear on the allies by closing the autobahn and other transport links, although twelve thousand tons of supplies were needed to keep West Berlin going each day. The RAF flew in supplies, with the Americans, round the clock – delivering some seven thousand tons per day, on average, from June 1948 to May 1949.

* * *

Alexander volunteered for Service with the RAF in 1941, having been a Driver in the Royal Army Service Corps since the start of the war. He was commissioned as a pilot in 1944 and at the end of hostilities Alexander was flying transport aircraft in the Middle East.

The Berlin Airlift was different from any other flying duties posting, in that aircraft flew continuously, day and night everyday and even in bad weather conditions, which would have put off most other air operations. There were days when the RAF were ordered to fly in fog which had already grounded the Americans. Alexander ended up with a nervous breakdown and he was invalided out of the Service with psychoneurosis. There then followed years of problems regarding the award of a War Pension in a case which would seem to have been of obvious merit. Only in 1952 was Alexander awarded a lump-sum payment for injuries said to be aggravated by the Service. This was difficult to understand at the time because he came out of the RAF suffering from nervous exhaustion. Perhaps it was felt that his condition was likely to improve. However, many more years later the award was increased to 40 per cent disability, but still remained for an aggravated condition.

Alexander undertook various jobs in engineering but problems always arose after a few years and none of his enterprises were successful. Alexander finally retired at sixty-two, an unhappy and frustrated man.

When I first met Alexander, he was being seen by a psychiatrist on a regular basis who was prescribing tranquilising medication which had calmed him down somewhat, making his life more tolerable. I found that, following admissions to Tyrwhitt House, it was possible to support a case for an appeal to increase his War Pension percentage further and to make his condition attributable to his Service and not merely aggravated by it. Exactly four years later, this was achieved with the support of the local representative from the RAF Association at the Tribunal, and Alexander came away with an award for 60 per cent disability attributable to his Service in the RAF.

In 1949 the North Atlantic Treaty Organisation was set up, involving the United States and Canada which was, in itself, an unprecedented step in world politics. This was to have a profound effect on Britain's involvement in Europe and lead to a major change in British foreign policy. One aspect of this change from pre-war isolationism and the defence of the Empire was seen in 1950, when Great Britain joined the Americans in the Korean War.

*   *   *

In November 1951, the King's Own Scottish Borderers faced a force of six thousand Chinese. **Bill Speakman** led repeated attacks at the enemy until he was overcome by his injuries. He was awarded the Victoria Cross and there were other awards for gallantry by the Regiment at this time. When the Gloucestershire Regiment was largely taken prisoner, only thirty-nine of their number reached the United Nations line and avoided capture or death on Hill 235, South of the Imjin River. In one battle the King's Shropshire Light Infantry found the Chinese had left 1,800 dead on the battlefield. In winter, severe weather went down to 16 degrees of frost. Both sides were evenly balanced so the war dragged on until July 1953. There were also naval operations to secure our supplies and to prevent the landforces from being outflanked by sea. The Korean War was the first campaign fought to uphold the authority of the United Nations; the prestige of the British Forces was enhanced but the nation's military and financial resources were stretched to the limit.

* * *

The most successful case I had from the Korean War was that of **Oliver**, who saw an article about the work of the Society in the Royal British Legion magazine. This gave him great comfort. He already had a 50 per cent War Pension for anxiety state and malnutrition after several appeals in the 1950s. So money was not Oliver's biggest problem

Oliver was called up in 1945, and served in Palestine, Greece and Egypt. On discharge he joined the Territorial Army and volunteered for Service with the Gloucestershire Regiment for eighteen months during the war in Korea. He was taken prisoner for two and a half years, suffered solitary confinement, and was badly shaken when the Americans bombed and attacked the prison camp. The strange thing was that, after all this, Oliver was discharged as being fit and able.

Oliver tried several jobs but found he could not settle. Eventually, he was taken on as a cleaner and gained promotion to laboratory technician. Oliver felt that the firm never gave him the further promotion he deserved after they found out about his war pension. When he suffered from a heart attack in his fifties, this led to early retirement. In conversation, it was revealed that Oliver had had six weeks in a mental hospital in 1957 which he said did little for him, but I feel this experience was probably instrumental in providing the appropriate reports which facilitated the granting of his War Pension.

When I met Oliver and his wife, they were worried because he could not come to terms with his mental problems; he was too nervous to consider coming to stay at Tyrwhitt House as this would have meant leaving his wife, on whom he felt dependent. I tried to have him seen by a consultant psychiatrist locally and I wrote to his GP. The doctor was unwilling to co-operate because he was of the opinion that Oliver did not have any serious mental problems. I doubt whether Oliver had told his doctor about his Korean experience because it made him embarrassed to talk about what he had suffered. The problem to be faced was that a consultant will normally only see a patient who has been referred by a GP. This is the way the system operates.

However, the Society did have links with military hospitals, and indeed, we knew many of the Service medical staff on a personal basis, as former colleagues. One argument put forward was that war Service pensioners are entitled to go to military hospitals in cases where the complaint affects their accepted disability, for which a pension had previously been awarded. In the circumstances, I referred Oliver's case to a consultant psychiatrist at Queen Elizabeth Military Hospital, Woolwich. As luck would have it, he was personally interested in the mental problems of ex-prisoners of war.

After a number of out-patient visits to Woolwich, Oliver improved out of all recognition. His self-confidence returned, and when I last saw him there was felt to be no need for further psychiatric treatment. I never discovered whether his GP ever found out about this. The War Pensions Agency were approached, and they paid all the expenses for Oliver and his wife to visit the Military Hospital.

Sadly, one of the most unsuccessful cases I ever had also came from the Korean War. **Benny** did labouring jobs until he was called up for National Service with the First Battalion the Gloucestershire Regiment in 1950. He was taken prisoner at the Battle of the Imjin, badly treated, starved and brainwashed. He was not released until two years later and came home for demobilisation after about four years' Service. At that time, National Service was normally for a period of two years.

Benny applied, unsuccessfully, for a War Pension for psycho-neurosis. It is possible that he was rejected because he had gone back to his old job working for his father, as a painter and decorator. But this work did not last and later he went to work in a factory. He drifted back into the building industry where he plastered ceilings until he began to deteriorate mentally in 1968. When I met Benny, he had been in and out of mental hospitals for some twenty years, never able to work, and relying on a fortnightly injection from the visiting Community Psychiatric Nurse. His parents had recently died and he was coping, with some difficulty, in what had been their Council house. His wife had left him fourteen years previously, obviously unable to cope with Benny's unstable condition. One of his few friends wrote and drew my attention to his plight.

We enrolled Benny as a client of the Society and admitted him to Tyrwhitt House for convalescence and a reconsideration of his case for a War Pension. After several appeals it appeared we were unlikely to succeed. I will never understand why – unless Benny's medical history contained information to which I was not privy. The Society's consultant psychiatrist was also surprised at the outcome. But his conclusion was that we could not reasonably expect to win every case, that we had a very satisfactory record of success, and to persist would only harm the Society's credibility. I then had the difficult task of explaining to Benny that it was the War Pensions Agency, and not the Society, who made the decisions regarding awards.

Sadly, Benny died of cancer two years later. I then had to explain to his relations why he never received the pension we all felt he deserved. As Benny had never been able to save any money at all, the Regiment agreed to pay for his funeral.

*   *   *

The prisoners of war in North Korea were treated in a manner similar to the Japanese prisoners in the Second World War, but in addition they had to contend with brainwashing or political indoctrination to remove from their minds the influence of capitalism. When the brain function has been sufficiently disturbed by fear, anger or excitement, the result is impaired judgement and the victim becomes very suggestible.

So it was that the North Koreans attempted to convert a few individuals, who would be used as ringleaders. Those who gave the wrong answers were put in solitary confinement. Those who toed the party line received better rations and prisoners were encouraged to write propaganda material for the Communists. Those who resisted were tortured, starved and confined in wooden boxes measuring five feet by three feet by two feet. Then there was physical punishment such as marching barefoot in 20 degrees of frost. The Communists regarded the prisoner of war camp as a battlefield. It is interesting that of the small number of Americans who refused to go home after the war, many were found to be of inadequate personality.

*   *   *

The trauma of war disability was illustrated by **Doug**, who came from South Wales. He was severely burned and lost a leg in Korea at the age of 21. Although provided with an artificial leg, he only worked on and off for the next fifteen years when his leg was further amputated. Doug was awarded a 90 per cent War Pension, never worked again, and was permanently on pain killers. Officially, he was not mentally ill, but we took him into Tyrwhitt House to give him some support after all his years of suffering and to give his caring wife a break which she deserved.

Jack was brought up by Dr Barnado's. As soon as he was able, he joined the Royal Navy on HMS *Ganges*. At the age of twenty, he was the youngest Petty Officer in the Service. Whilst serving with the Far Eastern Fleet, Jack served on warships; he was sunk on three occasions, including the *Repulse* and the *Hermes*. In 1949 he was in action again, this time on HMS *Amethyst* in the Yangtze, China. He served in ships offshore in the Korean War but had to be invalided out of the Service at the age of forty-five.

Jack never discusses how he received his injuries, but much of his inside consists of 'plastic plumbing' including stomach and bladder. Jack eventually received a War Pension for injuries to his abdomen, leg and pelvis. His mental problems stem from a variety of causes: he is afraid of losing his temper, and has found it for the best to live apart from his wife. He constantly harks back to the Navy which had been his whole life and, being obliged to leave at forty-five must have been a blow.

Jack exhibits an unsettled mind. One wonders whether it stems from his poor childhood, aggravated by stormy periods of active Service and injuries. He visits Tyrwhitt House regularly where he enjoys the company. The congenial atmosphere helps to calm him.

*   *   *

Almost immediately after the end of the Korean campaign British forces became involved in counter-insurgency operations in some widely dispersed parts of the Empire. British soldiers were already carrying out operations in Malaya shortly after the Second World War and the demands for independence were linked to Communist ideology. Right up to the 1960s the Army was seeking out 'CTs' (Communist Terrorists). During the war there had been some anti-Japanese operations carried out by the Malayan Communist Party under the banner of the MPAJA (Malayan Peoples' Anti-Japanese Army).

Independence was promised to Malaya after the defeat of the terrorists. Villages had to be fortified to deny food to the rebels in the jungle. At the height of the Malayan emergency there were 35,000 British soldiers stationed there. Five hundred soldiers were killed in jungle fighting and ambushes, but what defeated the rebels was the efficiency of the police and military intelligence Services. Eventually 6,000 CTs were killed.

The emergency in Kenya lasted from 1952 to 1960. There were originally three battalions of King's African Rifles. The Kikuyu tribe had grievances and Mau Mau terrorists used methods of ritualistic murder and terror tactics. At one stage, some 12,500 insurgents attacked isolated settlers and members of their own tribe who opposed them. By 1955, five British battalions and six of the King's African Rifles were employed on clearing operations working from Nairobi outwards, using cordon and search methods. There were also other gangs masquerading as Mau Mau who had to be tackled. By 1960, the emergency was over but not until six hundred British soldiers had been killed in operations.

The problems in Cyprus were even more intractable. In 1954 Britain decided that independence was not possible because of the strategic importance of the island. EOKA had only a few men but did have the support of the Greek Cypriots who formed the majority of the population. There followed a campaign of guerilla warfare terrorism and propaganda. An Emergency was declared in 1955 and EOKA resorted to ambush tactics in the Troodos Mountains.

A compromise was reached in 1959 after one hundred soldiers had been killed and British policy had alternated between negotiation and military action. In Malaya and Kenya military victories had been achieved before there was any talk of a settlement. All these campaigns showed the importance of joint military and civil police operations, and the painstaking use of gathering intelligence. The Army had to adapt to new methods of warfare which have been in use ever since. Traditional soldiering in these so-called minor campaigns was a thing of the past with no set-piece battles against a known and recognisable enemy. An incident could arise at any time, anywhere, and the local population was not always on the side of the troops sent out there to keep order. The men could never relax and found themselves constantly looking over their shoulders for trouble to break out. Nervous anxiety for some was never far away, and it never went away because, as soon as one incident ended, or a terrorist put behind bars, others would appear somewhere else. The military personnel very rarely had the satisfaction of meeting the enemy in a straightforward battle they were capable of winning outright.

*   *   *

A typical outcome of such situations is the case of **Roger**, now in his fifties. He joined his regiment in 1956 and served for three years in Cyprus. He was diagnosed as suffering from PTSD as a result of an ambush in which he saw a comrade being blown to pieces in his vehicle. This experience is something Roger can never forget. Before he was discharged, and before his illness was confirmed, the Army treated Roger like any other soldier. Whilst training on a rifle range, back in UK, his lack of concentration was such that he narrowly missed shooting an officer. This incident was to hasten Roger's discharge, by way of the Military Psychiatric Hospital, Netley.

After leaving the Army, Roger was a self-employed painter and decorator for many years. However, he gradually began to suffer from nightmares, blackouts and the partial paralysis of one side of his body. Towards the end of his working life some of his customers did not pay Roger the money they owed him for his work. He was too ill to follow up the bad debts and Roger ended up with substantial financial problems on top of everything else. The fact that he and his wife were trying to buy their house from the Council, on a mortgage, made matters worse. He applied for a War Pension in 1991 but all he received was the lowest possible percentage, which took the form of a lump sum. This only went towards paying off a few of their debts.

When I saw Roger he had already put in a deterioration claim and the War Pensions Welfare Officer referred him to me with a view to obtaining the support of the Society. We placed him in Tyrwhitt House where he was seen by the Society's Consultant Psychiatrist and eventually a substantial increase in his war pension was awarded. Now, feeling much brighter, Roger does household repairs for neighbours, but he is unable to work for long periods.

Edwin was in a reserved occupation during the war, volunteered for the Army in 1944, and saw Service in India and Singapore. At the start of the Malayan emergency in 1948, Edwin, as a trained soldier, was one of the first to be sent into the jungle in search of CTs, following a tip off. Enemy fire could come from any direction and so it was that Edwin was injured in the head from shrapnel and ricochet. Although he apparently recovered, Edwin developed a duodenal ulcer for which he received a War Pension on discharge. A few years later, he began to get blackouts and his GP diagnosed epilepsy in 1952.

Nevertheless Edwin went back to engineering for the next fifteen years until his employer found out about the epilepsy and dismissed him. In theory, anyone liable to suffer a fit could be in danger when operating a piece of machinery. In the event of an accident, the employer could be blamed for negligence.

As a result of this, the family moved from Scotland to the South of England. But his next job in engineering only lasted three years, until the firm found out about his War Pension. Edwin was sacked in 1970. From then onwards, living in a small village, Edwin thought it best to keep the epilepsy a secret even from his GP. He paid privately for his medication which came through the post and was very expensive.

Edwin built up an egg round, which developed into a shop, but he finally had to give this up when he began to suffer from heart trouble. I felt obliged to persuade him to make a clean breast of things to his doctor and he was eventually given a break at Tyrwhitt House. As a result of this his War Pension was increased and widened to include his mental problems. His medication for epilepsy is now provided by the NHS, which is a considerable saving to him. Edwin made friends at Tyrwhitt House and the whole problem of keeping his disabilities a closely guarded secret has shrunk into insignificance. He is a much happier man and his caring wife has benefitted from his improved condition.

Simon was born in 1946; his father had been a soldier and a prisoner of war. He followed his family round various army camps until he was old enough to join the Army himself. Simon saw active Service as a rifleman in Cyprus, Malaya and Borneo, where trouble first broke out in Brunei in December 1962. Troops were sent out to protect the Sultan whose authority was restored after a week. The following April, Indonesia threatened North Borneo and the SAS were sent in to work inside the villages gathering information. Later the Navy supported the action using helicopters and small ships. In December 1963 the Indonesians used their regular army but by August 1966 peace was agreed with Malaysia. Our Army lost 100 men in what was described as one of the most efficient uses of military forces in the world.

Simon was sent from Malaya to Borneo where he was wounded in the head. This resulted in his hearing being impaired. He came out of the Army in a state of shock, and there followed a sorry tale. Through the Regiment, he obtained work as a gun dog handler on a large estate but the noise of the shooting, naturally, upset him, although he enjoyed working with the animals. Simon became depressed and his marriage broke up. He tried his hand at looking after a boarding kennel for dogs, and he married again. However, his second wife left him when this business failed.

When I found Simon, he had been staying with his sister-in-law but the two had become incompatible. Simon was cracking up as a person, in the belief that no one could help him, that he could not relate personally to anyone, and that no one would employ him. In spite of this, he was smart and tidy in appearance living in a one-roomed flat paid for by Social Services. It was clear from the sister-in-law that his relatives could not stand him. His 30 per cent War Pension was used to buy food.

Simon needed long-term support, but as he was still in his forties he did not qualify for either of the two types of home run by the Society for elderly veterans or for short-term treatment. I therefore found him a place in a private home on the South Coast run by a former nursing colleague who had previously worked in the Society's homes. Simon was difficult to get on with and there were always problems with him, but everyone persevered. His War Pension was increased by the addition of psychoneurosis and eventually he was able to move out of the home into accommodation locally. Later I heard a rumour that Simon had married his landlady!

I received a letter from a GP in Kent asking me to visit a patient who was unemployed and who was suffering from nightmares. The doctor was himself an ex-Serviceman. I found **Albert**, a man in his fifties, living in a country cottage with his wife and a grown-up son. Albert had been a farm worker for most of his life but was made redundant in 1978 at the age of forty-five. He then did various jobs with agricultural machinery and finally he was a garage attendant until the business failed and Albert was again redundant at fifty-three. There was no work for him in that part of the country.

Albert went to Malaya as a soldier in 1952, straight from basic training. He became a dog handler in Singapore and travelled around Malaya, where dogs were used to track terrorists in the jungle. He described to me how he shot a terrorist during an ambush and how this and other horrific sights had worried him ever since. There was thus a possibility that Albert was suffering from PTSD, but his condition was certainly aggravated by the difficulty of coming to terms with the reality that he might never work again. I accordingly alerted the local War Pensions Welfare Officer who helped him complete an application for a War Pension.

There were other ways in which I was able to help and we decided that, if he proceeded with the case for a pension, I would get him into Tyrwhitt House with a view to having Albert seen by the Society's Chief Consultant who might be able to support his application. Secondly, neither he nor his wife had ever had a holiday, so far as they could remember. This had been a reflection of the nature of farm work and the low wages he had received when he had been working and living in a 'tied' cottage. The Society arranged a week's holiday at a guest house in Bournemouth, and paid for by the Not Forgotten Association. Understandably, the holiday had a wonderful effect on both of them, particularly the fact that someone cared.

Albert enjoyed Tyrwhitt House but the case for a War Pension was to run into difficulties, which were probably connected with the death of his mother in 1988. It was arguable that this latter event had more to do with his mental state than what had happened in Malaya some forty years previously; in addition there was the more recent factor of being unemployed. The fact that Albert was not medically discharged from the Army but that he left, to please his young wife, at the end of his military contract of Service, also did not help his case.

Albert continues to visit Tyrwhitt House and he is certainly more resigned to his circumstances. He has made many friends there. It is also possible that Albert and his wife could qualify for future holidays in Bournemouth.

Steve had served in the Special Air Service (SAS). He was a professional soldier in every sense of the word. After seven years in the Royal Navy after the war, Steve found he missed the excitement of Service life. He joined the Royal Corps of Signals and eventually became a sergeant in the Signals Squadron of the Parachute Brigade. This was to lead to active Service in the Suez campaign, Cyprus and then with the SAS to Malaya and Borneo.

Steve will say very little about what caused him to be invalided out as a result of head injuries, hiatus hernia and hearing loss. He was awarded a 30 per cent War Pension after discharge from Millbank Military Hospital in 1964.

Steve and his wife kept a public house for ten years, but I suspect that his wife did most of the work. Steve could not settle and frustration led him to take a job as a mine mechanic in Canada where he had a heart attack and has not worked since. His wife kept working by doing odd jobs cleaning and cooking after they settled in the South West in 1986. Two years later, Steve collapsed in the street.

When I saw him, Steve's memory for events early in his Service career was good, but poor for more recent happenings. He told me that when he came out of the Army he saw 'flashing lights' in his head, which came on at any time 'like explosions'. He was mobile but walked with a stick. He had spondylosis for which he received Mobility Allowance; he was also on insulin for diabetes. I found Steve to be very worried about many things he had seen in the SAS but which he would not discuss with me.

Clearly, Steve was in need of professional counselling from a military psychiatrist and I arranged for him to be seen by the Society's Chief Consultant at Tyrwhitt House. We were able to give him much needed reassurance and eventually his pension was substantially increased.

*   *   *

Steve had been at Suez where, in 1955, there was a significant turn of events for Britain as a result of the short-lived campaign to re-occupy the canal. The Army lost twenty-two men. The importance of this period was that Britain and France both tried to reassert their influence in the Middle East after the loss of India and Palestine. Russia threatened to support Egypt. Britain had financial problems, and was clearly overstretched. Her role as a superpower was clearly out-dated and decolonisation was to follow; Britain now relied on nuclear weapons rather than on armies in the field. British garrisons were run down but were to be capable of reinforcement. The phasing out of National Service made the Army's manpower problems even worse. The British Armed Forces were to be halved from 690,000 to 375,000 men.

*   *   *

Later campaigns were brutal, including events in Aden in 1966–7. Two clients who served there had totally different experiences. **Martin** was brought up in care. He joined the Army in 1964 and was soon posted from Germany to Aden during the emergency. He remembered one occasion when only one out of thirteen comrades on patrol returned to base. Martin was seriously affected by this and he was sent back to the UK where, in his words, 'everything blew up'. He was clearly out of his mind, at least temporarily. He was medically discharged and eventually awarded 20 per cent War Pension for pancreatitis, aggravated by his military Service.

When I saw Martin, he told me how he had appealed against the war pension percentage when he had been told that he would never work again. The appeal failed but this only made Martin more determined than ever to succeed in life. He married a nurse and has done self-employed painting and decorating ever since, with only occasional time off for illness. Martin was very determined to keep on working, which is why he declined to come to Tyrwhitt House. He certainly needed the convalescence in view of his state of nerves. He was on edge every time I visited him, and I was particularly careful what I said to him. Martin's problems have been physical, but there have been serious mental side-effects.

In declining to accept the addition of stress to Martin's pension, one suspects several factors were considered by the adjudicating authorities. Martin had an unsettled childhood and this was not the fault of the Army. Secondly, it came out that he was not actually on patrol in Aden and the horrors were relayed to him second hand. Nonetheless, the atmosphere in Aden must have been extremely tense and very likely to upset anyone with a weakness, whether physical or mental. In Martin's case the authorities accepted that his internal physical problems were made worse by his military Service. (I encountered comparable situations where Servicemen and women had served in Northern Ireland.) Martin should have received more help from the Society, but, whenever I called to see him by appointment, he was never there. He had been referred to me by the War Pensions Welfare Officer; Martin had been complaining to them at regular intervals, but there was nothing further they could do to help as all the appeals machinery had been used up. I tried calling without an appointment, but he was out at work. His wife told me he did not want to see me again.

Ron was a regular Serviceman who had had a chequered career before being granted a 70 per cent pension for PTSD after nine years' Service, first with the RAF then with the SAS. He served in Cyprus, Aden, Libya, Guinea and Guyana. Ron was taken prisoner by the Arabs in Aden who beat him up and imprisoned him in a hole in the ground. As a result of this he still has a hatred of Arabs generally.

After discharge, Ron was employed on Government communications work at GCHQ, Cheltenham; then at the Foreign Office; and finally at the British Council. All this lasted twelve years, after which he had a nervous breakdown and he was invalided out of the Civil Service with the promise of a pension at the age of sixty. In fact, in recent years Ron only received his War Pension on which to exist.

I met Ron in a Council flat in a country village where he lived with his wife on the income from his War Disability Pension. He was exempt from Council Tax and probably received Housing Benefit. There had been problems when Ron had felt suicidal and he had been known to disappear for days on end. He had some insight into his problems and was an out-patient at the RAF Hospital near Swindon. Ron was aware of his bad tempers and he told me about nightmares and flashbacks, which were usually connected with his Army experiences in Aden.

I was able to take preliminary steps towards getting him into Tyrwhitt House, to give his wife a break and to have him seen by the Society's Chief Consultant. I then alerted the War Pensions Officer to the need for a review of other possible allowances to supplement his War Pension. Then I gave him details of the local branch of the Regular Forces Employment Association (RFEA) in case he felt able to consider going back to work. It was recognised, however, that because he was now in his fifties, he might have to settle for doing voluntary work. Then there was the problem over the non-receipt of his Civil Service pension until the age of sixty. I felt that enquiries ought to be made regarding this, if it could be established, with medical evidence, that Ron was not going to be able to work again and earn a livelihood. I had been involved before in cases where disabled ex-Servicemen drew a modified Army Pension before the age of fifty-five on medical advice that they would never work again. As his military Service had been on Short Service engagements, it was possible that these years had been added to his Civil Service pension, because he never qualified for a Service Pension.

Clients such as Ron can take up much time, but the reward comes when, after a long battle, the client might win his case and enjoy an improved standard of living or a better state of health in mind or body.

Two years later, I found that Ron had never gone to Tyrwhitt House. He was reluctant to leave his wife, and his physical problems increased. For some individuals, a reluctance to stay away from home was based on a selfish fear that this person's wife, on whom the individual depended, would no longer be there on his return; or it might be based on a plain fear of the unknown, or a feeling of inadequacy facing other people who were not familiar. Feelings of agoraphobia are not uncommon. Going away from home, for the first time in years, and alone, can sometimes bring back all the worst memories of joining the Army again. However, it is hoped that Ron and his wife might be able to take a week's holiday in Bournemouth, sponsored by the Not Forgotten Association.

7

Help is Available – but is it enough?

As we have seen in the preceding chapters, tens of thousands of ex-Servicemen and women are today suffering from mental problems which are the legacy of their traumatic experiences in the Armed Forces. Before bringing the story up to date, and dealing with the military exploits of the most recent times, it is appropriate to consider what is currently available to ex-Servicemen and women by way of help. Basically, three kinds of help may be sought: care, cash or a cure.

In the past, there were never sufficient resources to match the requirements of all these people, nor will there ever be. This is because we are dealing with an enormous social and medical need against the background of a National Health Service which, half a century after its inception, has been proved incapable of resolving all the health problems of all the people for the whole of their lengthening lives. Albeit with the best will in the world of those who work in the NHS. My own personal experience as an administrator tells me that, when we spend public money on schools, hospitals or social Services, we can never hope to solve all the problems of everyone who uses the Service. This is because we are dealing with real people in a real world of scientific research, technological development and increasing aspirations and demands on the part of the public.

Furthermore, merely throwing money at a social problem seldom provides the complete answer; indeed it may well reveal even greater problems than were first envisaged. A real danger is that those who strive to help the sick or the deprived, or to educate our children, may lose heart when they see that as fast as certain matters are improved, other difficulties arise. And, indeed, the sheer volume of work seems to increase rather than subside, or, at best, stand still.

The population, including the sick and the able-bodied, continues to live longer. Recall the old song, 'Old Soldiers Never Die' – it's true! Every year that goes by, the same faces reappear in the convalescent homes I visit. There are of course those who have departed, but in the main the number of clients always increases as new cases are discovered. Though it has to be said that some inmates are encountering mental illness for the first time, often connected with increasing old age, and not always connected with previous Service life. Given this situation it is highly likely that thousands suffered in silence. Although they may have been supported by relatives, many were too dignified to complain, and they died in pain. The generation born before the Second World War never expected there to be a comprehensive Health Service to which they could turn without having to pay. They counted the cost in money before calling the doctor in the 1930s, while religious conviction led some to accept suffering as an act of selflessness. The Welfare State was to lead to a complete change of attitude on the part of a large section of the population. When the Labour Government took office in 1945, with a land-slide majority, expectations ran high. But their children were to find out that welfare handouts have to be paid for by someone, in the long run. Some of the political gossip of the 1920s and 1930s had led to a belief in an endless source of wealth which would become available once all the idle rich were made to pay for every-thing. It was never understood just how few really rich people there were compared with the millions of deserving poor, and how getting rid of the very rich would do little to help the needy multitudes.

No social or medical problem is capable of total resolution, for, as quickly as one area of improvement is achieved, so standards and expectations will rise. This need not necessarily be based on greed, but on the optimism of the human spirit. Medical research has prolonged the lives of millions; it has also enabled millions to live in reasonable comfort in their bodies giving them hope, greater expectations and increased demands for attention, medication and a cure. The problems of the mentally disabled, be they ex-Service or civilian, are not going to disappear. The problems are deep-rooted within our society, and cannot be totally resolved by financial handouts.

Before the arrival of the Welfare State, it was customary for the family or the extended family to resolve its own social problems. Aged relatives were invariably looked after in this way, as were the unemployed and the chronically sick and disabled. But it was a different story when it came to the mentally ill. Every large area of population in the country had a Victorian mental hospital, set in rural surroundings, designed to provide seclusion, occupation and some treatment for patients who were expected to remain there for most of their lives. Some were visited regularly by relatives, but many were simply abandoned through feelings of shame, or just neglect. Some patients were frustrated by their surroundings and probably ought not to have been there indefinitely in the first place. Others became resigned to their fate and were perhaps happier than they might have been elsewhere.

Advances in medical science in the form of tranquilising drugs made it possible for many hospital patients to be considered for release into the community, with the consequent run-down of these large institutions and the transfer of some of the patients' problems to the world outside. I was to meet clients who were still in long-stay hospitals; others were successfully released into the community; and yet others who were in the community at large, but who would have been better looked after in hospital. It is of course distressing to hear of mental patients being released only to commit serious crimes or cause harm to themselves. Some relevant case histories will be detailed in chapter 9.

For families, the primary problem is the stigma of mental illness. Those who last century put away an eccentric relative would nowadays expect the NHS to place him or her in a residential home or a secure hospital, and to pay the fees. Some of my clients chose to live alone so as not to allow their presence or their reputation to react unfavourably on their relatives. Insanity has always been thought to run in families, although the mental illness of the majority of my patients had nothing at all to do with their relatives – it was brought about by stress encountered under Service conditions.

I have referred to 'cash' or money as a prime need of the disabled ex-Service community. Fortunately, they have access to four principal sources, two from the State, two from ex-Service organisations.[1] Clients usually qualify for Social Security benefits, the same as the rest of the population, but there have been a few exceptional cases, notably with regard to the State Retirement Pension, when the requisite minimum number of contributions for a full pension has not been made. This can apply to the chronically sick in hospital, those out of the country and, in a few cases, those too ill to bother to pay contributions or to claim exemption, or to claim the pension on reaching the right age.

*   *   *

Stephen was an ex-infantry officer, and his wife was in the WAAF during the war, when I discovered them, in 1986, at the request of the War Pensions Officer. They were in serious financial trouble. Stephen was just 65 and had been too ill to work for years; his wife had had to give up working recently, aged 62. Stephen had a small War Pension for psychoneurosis, which had not been reviewed for about thirty years. He had not claimed sickness benefit until his wife could no longer earn a living for both of them. Stephen was never accustomed to thinking about money and, as with many of my clients, his illness seemed to make him totally disregard its value. They were short of clothing, basic household running repairs were overdue and there was an eventual realisation that Stephen would never qualify for a full OAP because of the long periods when he had been too mentally ill to claim cover for periods when National Insurance contributions were not being made. The War Pensions Officer took charge of this latter problem and she obtained for Stephen the best OAP that could be provided under the Rules, making every possible claim for exceptional circumstances. In fact, we arranged to put Stephen on Invalidity Benefit, with the support of his GP. He received this benefit until he was seventy, which he was allowed to do as it was to Stephen's financial advantage.

Stephen's Regiment and the Royal Air Force Benevolent Fund gave substantial help towards resolving his various financial difficulties. Stephen was admitted to Tyrwhitt House, and he has benefitted from a stay there every year since, thus giving his wife a much needed break from caring for him. Stephen's grown-up children were very supportive but, with their own problems, there was no way that they could have carried out the 'rescue operation' their parents so badly needed. For many years the children were kept in ignorance of their parents' plight.

*   *   *

Those who qualify for the State Retirement Pension may also find that they are entitled to Income Support. Nowadays, the OAP is not a fixed amount for everyone and, apart for having worked for 35 years with contributions paid, the nominal period of working life to qualify for the full basic OAP, there have been additions for having earned more and contributed more than the basic rate, since 1961. Those who have a pension based on a working life less than 35 years, for whatever reason, may receive a proportionally lower State Pension. This can entitle the pensioner to claim Income Support.

*   *   *

Alan served in the Army as an infantry soldier in Burma. On the way out, by troopship, he landed in South Africa where he met the young lady whom he married after the war, when he was invalided out of the Army. He received a small War Pension for depression.

Cedarwood Guest House, 1991. The Author, Joan and Fred Pritchard (Proprietors). Seated, Alan and his wife from Zimbabwe.

They went to live in Rhodesia but eventually international trade sanctions against the 'Smith Regime' led Alan to become un-employed. He eventually got a job on Rhodesian Railways until the Zimbabwe Government made him retire at sixty and there was no OAP due. In the meantime, his War Pension had been re-duced to below 20 per cent, so Alan was given a lump sum and paid off.

Alan and his wife came to London with no money, no savings, no job, no pension and no Social Security allowances. They slept on the floor and his old regiment gave them money to buy some blankets. I found them in a small flat in London, surviving on an extremely small OAP (because of all the years when he was working outside the country), topped up by some Income support. The OAP was based on Alan's work before he was called up, his period in the British Army, and finally the years from the age of 60 to 65 when he did menial jobs.

One of the nicest things I ever did was to send them both on a week's holiday to Bournemouth, paid for by the Not Forgotten Association. This was the first time anyone had shown this couple that they cared. Later on, we persuaded Alan to go to Tyrwhitt House, although at first he was reluctant to leave his wife, after all that they had been through together. His stay was financially assisted by the Local Authority; this, in itself, was a major break-through when many authorities were pleading that they had no money to pay for clients to have periods of convalescence. While he was there, the Society's Chief Consultant supported a case to have his War Pension restored and increased. This was to give them a better standard of living and it also meant that subsequent periods at Tyrwhitt House would be paid for by the War Pensions Agency.

*   *   *

Income Support is a Social Security benefit available to all adults, whether working or not, whose income is below a certain level. Persons on a low income may also qualify for help with the rent in the form of Housing Benefit, which is paid by the Local Authority. Council Tax payments may also be reduced.

In addition, there are allowances for sick, injured or disabled people of all ages and about half of my clients who were not working might qualify for some of these. Incapacity Benefit replaced Sickness and Invalidity Benefits from April 1995. Some clients qualified for Attendance Allowance at the higher or lower rate. The Disability Living Allowance replaced the Mobility allowance a few years ago, and there is also a care component for those who need personal care through illness or disability.

About half of my clients were in receipt of War Disability Pensions. Nowadays these are administered by the War Pensions Agency, and may be awarded to anyone disabled or injured as a result of Service in the British Armed Forces or the Polish Forces under British command, or as a civilian, or a merchant seaman injured by enemy action, or the widow or dependant of such a person.

A War Disablement Pension is assessed on a percentage scale from 20 to 100 per cent. An assessment below 20 per cent may lead to a lump-sum payment. War Pensions and lump sums are tax free, and there are other allowances in the scheme, mainly similar to the civilian benefits. These include the War Pensioner's Mobility Supplement, the Constant Attendance Allowance, as well as allowances for a lowered standard of occupation, or for someone who will never be able to work at all in future. A war pensioner who dies as a result of his accepted disability may qualify for funeral expenses and a pension for his widow. He may be able to receive priority treatment for the same disability at a military hospital or an NHS hospital. In recent years, this latter recommendation has run into some difficulties because civilian hospital staff have had no experience of handling the casualties of war, the mentally ill, or do not understand the arrangements for priority treatment for veterans.

The rate of payment for a War Disability Pension has been about £1 a week for every percentage of disability assessed; this is a very rough guide, but useful as an explanatory tool when trying to encourage someone to apply for the first time.

All applications are considered by the War Pensions Agency at Norcross, Blackpool FY5 3WP; there is a War Pensions telephone help line on 01253 858 858. The Service is regionalised and the officers of the War Pensioners' Welfare Service are based at each of some thirty regional welfare offices throughout the British Isles. Anyone who feels that he or she has a case for a War Pension may seek advice. All who already have a Pension may be helped regarding additional allowances, or if it is felt that the percentage disability assessed is not a correct indication of their present condition. I have always found War Pensions Welfare Officers, without exception, to be highly motivated and extremely competent and devoted to their clients. They really do help the claimant obtain the best out of the system. This is a truly remarkable state of affairs when one considers that hardly any of the Welfare Officers today ever actually served in the Forces themselves.

Applicants must complete the appropriate form which requires details of the individual's GP and any hospital treatment received. This information is followed up and referred to a medical team at Norcross, together with the client's Service medical record. For cases opened many years after leaving the Forces the onus is on the applicant to show that his disability was caused by Service conditions and not by something he already had before he joined, or has acquired since he left. There are times when the most obvious case for an award is rejected. But there is always the opportunity to appeal, in which case additional evidence, perhaps provided by the Society's Chief Consultant, might be of help.

Should the Appeal fail, the case can be referred to a Tribunal consisting of three independent persons to whom the applicant may appeal directly, accompanied by a friend who may be an officer from one of the ex-Service organisations. I attended tribunals in London, Birmingham, Cardiff and Exeter. Some were against the non-award of a pension (an entitlement appeal); some were where it was thought the percentage pension awarded was too low (an assessment appeal). Any pension paid, or any higher rate awarded, is only paid from the date the original application or appeal was made. Arrears are not normally paid right back to the Second World War, for instance![2]

In cases where I attended and gave evidence, we were successful far more often than not. With only one exception, I found the tribunals extremely helpful, giving the client and the Society's officer every opportunity to explain our case.

The person addressing the meeting does not have to stand and formalities are kept to a minimum. However, an important aspect is that the client's case is always fully documented, making use of a written Statement of Case. The applicant is then encouraged to explain his position to the Tribunal in his own words. Whilst it is important not to put words into his mouth, his feelings can be explored by asking him the right questions. In order to do this, some preparation is essential with the applicant himself. However, the most important factor is that whoever supports him knows him well, enjoys his confidence, and that these aspects are conveyed to the Board. In one or two cases, clients were unable to speak at all – such was the emotion, fear or physical disability.

War Pensions Headquarters, Norcross, or the Tribunal itself, may call for additional medical evidence by referring the case to another consultant. In such cases the Society's support was invaluable. It was sometimes necessary to ask the consultant to visit the client in his own home, if travelling was a serious problem. There were patients who suffered from severe physical disabilities, and some suffered from extreme agoraphobia. Then there might be problems associated with meeting someone strange. I was always available to take a patient to see a consultant or to be there if the consultant visited the patient in his own home. I had to judge the situation when deciding, with the doctor, whether to be present all the time, or to see the consultant separately before or after the client. Whatever the outcome, clients invariably felt that the Society had promoted their best interests. And that was its own reward. Society Officers did not represent the Army, War Pensions, or the NHS. Yet their wide knowledge of Service conditions, and of the problems of mental illness and its consequences, enabled them to promote the best interests of clients.

However, should anyone be successful in obtaining a War Pension, which is tax free, their financial success might be reduced by their Local Authority, depending on where they happen to live. Local means-tested benefits, such as Income Support or Housing Benefit, may be affected depending on whether or not the Council takes the War Pension into account when assessing, say, a Council house rent.

Alfred Williams was a pre-war regular soldier who went to France with the BEF in 1939 and escaped via Dunkirk. He landed in Normandy with the first wave of British soldiers in 1944, only to be evacuated home after the Battle of Caen. He served his full engagement, coming out in 1946, and was eventually awarded a 90 per cent War Pension for severe hearing loss and the effects of

Alfred Williams and his wife during the war, and afterwards.

battle fatigue. He had to pay the full rent of £70 a week for a tiny Council bungalow with a warden on the site. If his pension had been disregarded in terms of the assessment made, his rent would have been of the order of £40. These positions and rules are often historical and would appear to be unaffected by local party politics. It is assumed that as time passes the financial advantage to the council resulting from not disregarding War Pensions will become smaller and smaller, as fewer war pensioners continue to survive. (The case of Ned on p. 56 illustrates this aspect.) In fact Alfred was prominent in a local campaign resulting in total victory for the pensioners in 1998.

It can happen that a client attending a Tribunal is represented by one of the large national organisations, and that the official only meets the applicant in person just prior to commencement of the proceedings. If a client is nervous or apprehensive, or simply unable to express himself at all, the lack of personal contact can have an adverse effect on the result. Some ex-Servicemen remarked that they felt as if they were themselves on trial. But this problem must be understood in the context of handling many cases, and where the representative has to travel long distances in order to cover a large territory.

The War Pensions process takes many months to achieve a result. Should it be necessary to go to a tribunal then the Statement of Case is prepared by Norcross. It will contain detailed extracts from every available piece of medical evidence held on the client since the day he joined the Forces, and afterwards. The system used to be notoriously slow. It was not uncommon for a client to remark that they would be dead before any outcome was reached. However, in recent years considerable progress has been made in speeding up the process. Indeed, the War Pensions Agency has been the proud possessor of the Government's 'Chartermark' for running an efficient, caring Service. This was well deserved at the time.

In spite of protests criticising the lack of help available from public sources, in one or two cases people have done surprisingly well out of combining what is available from the DSS, the Local Authority, War Pensions and the ex-Service charities.

*   *   *

Pat is ex-ATS, physically handicapped and in a wheelchair. **Tom** has nervous anxiety. He is physically fit but unreliable in behaviour, she is very sensible. They claim every possible allowance and live in comparative luxury. He has a high percentage War Pension, and she gets Disability Living Allowance and Attendance Allowance for looking after him. They both receive State retirement pensions. They live in a modern flat for the disabled and are known to every relevant charity. Lump sums have come from the regiments and from holiday charities. Tom benefits regularly from Tyrwhitt House.

One day, I had just arrived and I was asked to curtail the visit promptly. A taxi arrived, which obviously could not be kept waiting. It had come to take them shopping, and they were dressed up to the nines! I had mixed feelings about them being able to afford this sort of life – all out of the combined resources of the welfare Services. But when they sometimes complained that they did not receive Income Support, I took their complaints with a pinch of salt. All in all, they did very well: they received free holidays from a local association for the disabled, and a new carpet was purchased with help from one of the regimental benevolent funds.

*   *   *

A major source of benevolence lies with what are known as the Voluntary Bodies or charitable institutions. Some are specifically for ex-Service personnel, others are for specific types of medical need. The *Charities Digest*, published annually since 1882 by the Family Welfare Association (see Appendix), analyses all the hundreds of organisations of which it is aware.

For our purposes, there are two main categories, the medical and the military. Medical charities tend to be divided between research foundations and those that provide treatment, relief and care. There is a whole range of organisations devoted to the mentally handicapped. Many of the conditions from which my clients were suffering have an appropriate society; these include, Age Research, Alzheimer's Disease, Arthritis, Asthma, Epilepsy, the Chest Heart and Stroke Association, Tinnitis, Tropical Diseases, Schizophrenia, Parkinson's Disease – to name but a few of them. Contact with such a group brings with it the support of others with similar problems, news of research into treatment, and that rare commodity – the dispensation of hope and cheerfulness. The *Charities Digest* also contains details of hostels, housing associations, hospitals and residential homes. It is always useful to check to see whether a client's earlier career or background might entitle him to consideration for help from another charity – a trade union, for instance.

Being an ex-Service person opens up an entirely separate range of help from the military charities. However, being mentally disabled can promptly close some of their doors again, rather quickly. I shall expound on this in the coming pages.

One of the threads running through this book is the realisation and recognition by the military authorities that mental illness is not a form of cowardice and that, properly treated, thousands of mental casualties were able to be successfully rehabilitated, sometimes even to return to the front line. Like the civilian charities, the military ones can be divided into several categories. The main division is between those that hold funds which they disburse to the needy, and those that offer care of some kind – long-term, short-term or for a particular category of client or handicap.

Any account of the ex-Service organisations must commence with the Royal British Legion, which offers a wide range of support to the ex-Service community. This includes accommodation, permanent and temporary. There are three homes providing two-week breaks under the Rest and Convalescence scheme, in addition to the residential homes.

The Legion provides some employment, there are grants and loans for business enterprise, small gifts for the relief of hardship, a medical rehabilitation centre and, recently, an expansion of training facilities preparing ex-Service personnel for civilian employment. Then there are the ubiquitous RBL clubs, which are centres of social life and fund-raising. There are over 3,000 RBL branches. A branch may function without a club, but every club must have a parent branch. In addition, the Legion is involved in War Pension cases where experienced staff at Headquarters assist deserving clients to prepare and present their case. Around 50,000 pensions applications are handled each year and success, measured in terms of annual pensions and lump sums granted, amounted to over £18 million in 1993. Pension claims are on the increase and all who served in the Second World War are now old-age pensioners. Many are experiencing the combined effects of old age and lack of employment.

The Churchill Centre for Rehabilitation and Assessment at the Royal British Legion Village, Maidstone, provides treatment for ex-Servicemen unable to obtain such facilities elsewhere for arthritis, strokes, amputations and war injuries. The treatment available includes hydrotherapy, physiotherapy and occupational therapy. The treatment regimes are normally for about two weeks, and residential accommodation is available. Also at RBL Village is the RBL Industries Vocational Assessment Centre which aims to identify and develop the abilities of disabled ex-Servicemen and women in the difficult transition to appropriate civilian work roles.

There is an Officers' Benevolent Department, including an employment centre, placing about 500 ex-officers annually in suitable jobs. Indeed, I obtained my position as Regional Welfare Officer with the ex-Services Mental Welfare Society by this means, and I have always felt indebted to them for this.

The RBL Training College at Tidworth was opened in July 1994 to provide courses in business management, computer studies, child care, and starting a small business from scratch. This is a purpose-built college with facilities that can be used for large meetings, accommodation and entertainment. Then there is the Small Business Advisory Service and Loan Scheme. This is located at Tidworth but is part of the Legion's Welfare Division not the Training College.

About sixteen hundred ex-Servicemen are employed by the RBL Attendants' Company as security guards and car park inspectors. It will change its name to Legion Security on 1 October 1998. Royal British Legion Industries has a workshop at RBL Village, Maidstone, doing printing, stores distribution and some manual work making wooden pallets and road signs. There is a craft centre for the disabled, some of whom work from home, and the produce is sold in their clubs. Then there is the world famous poppy factory which produces over thirty million poppies a year.

As the County of Kent was in my area, I enjoyed good relations with the staff at Royal British Legion Village, Maidstone. An ex-RAF client benefitted from hydrotherapy treatment, for injuries to his back, at the Churchill Centre. This was an important achievement because there were no comparable facilities anywhere near where the client lived. (This was Adrian, mentioned on p. 000.)

I was able to help a couple obtain work in the stores at RBL Village, and a cottage to live in. In another case, an ex-Naval rating, whom the Legion and the Society had helped obtain a much deserved War Pension, developed problems even though his work in the stores was not arduous. The nervous anxiety from which he suffered was aggravated by the nature of his job. Being a caring employer, the Legion kept on reducing his hours until in the end he had to be given early retirement. Even then, permission was granted for him to retain the tenancy of his tied cottage.

Henry was invalided out of the RAF suffering from nervous anxiety. He was awarded a War Pension which he never drew because of the stigma of being mentally ill. He felt that it would prevent him from obtaining work and there were justifiable grounds for this attitude. Henry tried his hand at office cleaning, and working in a road-mending gang. He always did menial, low-paid work. I found him to be totally unreliable and probably schizophrenic in that much of his speech was nonsense. He lived alone in squalor, but I managed to obtain a vacancy for him at Tyrwhitt House. Henry's father drove him there, and Henry promptly disappeared! Later, we arranged an employment interview at RBL Village, but he never showed up. Subsequently no job could be located for him. Even his father eventually accepted that we were not going to be able to help. At the same time, however, the father refused to have his son live at home. He told me that Henry would embarrass them in front of the neighbours.

The Royal British Legion inevitably suffers from two disadvantages. The first is to do with the sheer size of the organisation, the second is its dependence upon thousands of volunteers who serve on committees, visit welfare cases and raise money. The Service delivered to the ex-Service person at the receiving end is bound to vary according to the volunteers involved in each case. Every Legion branch is expected to have a welfare committee whose members may disburse small amounts raised locally, to help needy ex-Servicemen, or refer the case to the County Field Officer (CFO). In one case I tried unsuccessfully to persuade someone from the Legion to visit a poor old sailor abandoned by his family in a home on the South Coast, a man who had nothing at all, not even the pocket money supposedly allowed by the DSS. (This was Ben, mentioned on p. 77.) By comparison, when a relative of mine had a civilian accident at work, being ex-Forces and a Legion member, he received every consideration including a free holiday in a convalescent home, with pocket money provided!

Some of the problems I met were undoubtedly to do with the reactions of the public at large towards mental illness. I know that admission to Legion homes, both permanent and temporary, is not encouraged where there is a history of mental problems. This is perfectly understandable when one difficult resident can make life quite impossible for the others, besides taking up an unfair proportion of attention from the staff. This is all the more unfortunate when many of those known to me, and in need of convalescence, are living on their own anyway, having been abandoned by their relatives because of their illness. Some of these clients who were in Legion and other charitable homes were successfully transferred to the Society's permanent home at Kingswood Grange; some would have otherwise ended up in NHS institutions for the mentally incurable. In all fairness, it has to be said that the Legion would rarely discharge someone from a home on the grounds of mental illness alone, unless into the care of the Society. A few, successfully and unsuccessfully, have been taken into the Care in the Community philosophy.

I have met RBL Welfare Committee members who told me that, now that the Welfare State cared for everyone, there was really very little for them to do. This is not a view shared by RBL Headquarters. I suspect that the mentally disabled ex-Servicemen were simply not known to the Welfare Committee because these individuals had not asked for help; people tended to avoid them and they were probably not club members anyway. Some of my clients were obliged to keep away from clubs because they were not permitted to mix alcohol with their medication, and found it difficult to be in the company of others enjoying a drink.

In recent years the Royal British Legion has greatly strengthened its local organisation by the appointment of full-time County Field Officers whose task it is to oversee both welfare and fund-raising activities in their area. By giving full support to the volunteers and using the latest information technology these new officers have a greater control over what is spent and how funds are raised.

Naturally, the Legion and all the other charities are apprehensive about the impact of the National Lottery on fund-raising. Even before the Lottery commenced in late 1994, they were digging into their resources in order to maintain Services. In a recent year, £23 million was spent on benevolent work but only £13 million came in from the Poppy Appeal. Not withstanding legacies, investments, and some business enterprise funds, the Legion had to dip into £2.8 million of its reserves in order to balance the books. Clearly this cannot go on indefinitely.

In 1994 the start of the National Lottery was delayed until after Remembrance Sunday in order not to clash with the annual Poppy Appeal. All the charities must now compete with the Lottery, in addition to increased competition from the now commercially organised children's and animal charities. It is possible that in the future the charities discussed here could become dependent on donations from the National Lottery, with persons outside these organisations effectively deciding matters of policy and expenditure. Nonetheless, the Legion has reason to be pleased with the Poppy Appeal results in more recent years: £16.4 million in 1995, £16.1 million in 1996 and £17.3 million in 1997.

The Soldiers' Sailors' and Airmen's Families Association (SSAFA) amalgamated with The Forces Help Society and Lord Roberts Workshops (FHS/LRW) on 1 January 1997. The combined charity with the new name of The Soldiers, Sailors, Airmen and Families Association-Forces Help (SSAFA Forces Help):[5]

> Offers friendship; practical welfare advice and support; financial advice and support, training and employment for the disabled residential care; short stay accommodation and a Housing Advisory Service, to serving and ex-Service men and women and their families in need.
>
> Employs over 400 health professionals and social work staff to assist Service families at Service establishments around the world. (This work is carried out under contract from the MOD).

Advises the MOD on statutory obligations for providing health care; health education and social Services.

Acts as Agents of the Service, Regimental and other funds in bringing relief to those in financial need.

SSAFA Forces Help is a highly organised befriending Service based on a county organisation.The Association prides itself on prompt action wherever the need arises. There are some seven thousand voluntary workers in the British Isles who help more than 85,000 people annually. There are more than 500 In-Service volunteers on HM Forces bases around the world.

Because I did not specifically deal with families I did not often encounter SSAFA. There was the occasional visit to a client who happened to be in one of their residential homes and, exceptionally, this individual might visit Tyrwhitt House to see our Chief Consultant. Where I discovered that an ex-Service family needed help, which I could not give because I was not based near where the family lived, I would attend to the man's medical needs and perhaps sort out his pension, and SSAFA would be invited to attend to the more familial problems.

*   *   *

Noel wrote to me from his married quarters at Tidworth when he was about to be discharged from the Army with a nervous breakdown. Clare, his wife, was concerned as to where they were going to live; they could not stay on in Army quarters indefinitely and there was no civilian work for him on Salisbury Plain. Noel was a Lance Corporal, in his twenties, and there were three children. I visited them only to find that every suggestion I made regarding finding somewhere to live or a job was rejected by one or both of them. I investigated the possibility of a disability pension, as they were both very vague as to whether one had been applied for. When I returned, as promised, a few months later, they had left. Forwarding addresses were at relatives where I knew they had adamantly stated they were not going to be.

Two years later, Clare wrote to me from the South Coast to say that Noel's condition had worsened and that he had left her and the children. I called within a couple of days to find her living in a pleasant new home which they were buying from a

Housing Association. Noel had got a low percentage War Pension and he had been working as a security guard. However, he was becoming increasingly unable to cope with his life, and there were sizeable debts. She did not know where he was, or what to do next.

I was able to request a visit from a SSAFA representative who arrived within twenty-four hours. Rent arrears, allowances and entitlements were all sorted out. I traced Noel through his War Pensions Officer and we had him admitted to the Psychiatric Wing at Royal Naval Hospital, Haslar, for treatment. I then took steps towards encouraging his employer to take him back. Noel was able to return home, feeling much better. This case is a good example of two welfare agencies working well together.

Specialised ex-Service charities include St. Dunstan's (originally for those blinded in war), the British Limbless ex-Servicemen's Association, the Royal Star and Garter Home (for the physically disabled), and the Royal Hospital, Chelsea – all of which offer residential accommodation. The Not Forgotten Association really does live up to its name, and there are other charities which dispense goodwill, gifts, holidays and good cheer to ex-Service folk who otherwise would lead lonely lives. The Army Benevolent Fund gives grants to other charities who help ex-Servicemen; the Fund also helps the individual Regiment deal with cases which would be beyond its own resources.

*   *   *

In nearly every case of hardship I have encountered it has been the Regiment which has provided any money required. The ex-Services Mental Welfare Society can rarely pay out cash from its funds, all of which are required to finance the nursing homes, meeting shortfalls, together with covering the cost of the medical, nursing and welfare visiting Services which it carries out. Monies paid out to relieve urgent hardship cases is recovered, wherever possible, from the man's Regiment.

The attitude of the Regiments, understandably, varies enormously. Some no longer exist, anyway. Most of them have financial problems nowadays as they were originally financed on the basis of one day's pay subscribed by every serving officer and man. In the days when millions were conscripted, the sums of money thus collected were considerable. Since the end of the Second World War, the number of serving soldiers has declined substantially with every review of Defence Expenditure. The millions who contributed and fought in the war are now old-age pensioners, some of whom face illness and hardship. The Service charities in recent years have been hit by competitive fund-raising from animal, children's, conservation and medical charities. Some have had to decide, with the greatest regret, to concentrate their work on their ex-long Service Regular Forces members.

The amount of help an ex-Serviceman might receive will depend on which Regiment or Corps the individual served in. Size of the regiment does not appear to directly affect the situation as some of the smaller organisations, in my experience, can be more generous than the holders of very large funds. The Ministry of Defence (Personal Services, Army. PS4(A)) publishes a list of every known regimental fund. In the main, regimental secretaries rely on other welfare agencies to visit and make recommendations for the award of grant aid. For this purpose alone, activities such as mine could have been fully justified but, in fact, obtaining cash grants from the regiments was only a small part of my work. The lack of local representatives was, I felt, unfortunate as the larger regiments and corps could have improved their local representation and not relied so much on others to visit their claimants. Then there were other aspects of welfare work which the regiments seldom covered.

I remember a very elderly couple who badly needed support in order to arrange a move from a house on top of a steep hill to somewhere more convenient, nearer the town. The Regiment never did succeed in sending anyone to befriend them and help them move.

Generally, regiments prefer to make a grant for a single purpose on a 'once and for all' basis. They very rarely would present the applicant with a cheque followed by a suggestion that he might continue to apply for more. In my time, one or two clients received about £5 a week to relieve a permanent hardship with perhaps an inadequate OAP, no War Pension and inescapably high living costs, perhaps caused by sub-standard accommodation with high fuel bills. Army charities hardly ever made such payments. The charities were limited by the Government in the size of regular grants that could be made in this way – possibly for fear of under-mining the Social Security system! Regiments nearly always paid travelling expenses for a client coming to Tyrwhitt House and who was not on a War Pension with expenses paid by Norcross, as part of the approved treatment for the individual's accepted disability. Regiments also paid for visits to a sick relative or to the grave of a former soldier. We always handed out a few pounds to our needy clients at Christmas and much of this was refunded to the Society by the regiments afterwards.

A client who got himself into debt or encountered unforeseen circumstances might be helped, but this was an exercise which was seldom repeated. There were one or two who relied on the Society to obtain grants from the Regiment towards keeping an elderly car on the road and this made all the difference to our friends' quality of life. One had to apply each time, but grants might be awarded for vehicle insurance, road tax or the MOT test if the application was strongly recommended. A client living alone because of his mental illness, or afraid to go on public trans-port, could receive help towards keeping his car on the road, a godsend or a lifeline. A claustrophobia sufferer would benefit enormously from having the means of getting out of his small flat into the countryside. There were some whose condition was improved psychologically from being able to convey sick or infirm neighbours to the doctor or to a day centre; this made them feel wanted and gave them some status in their local community, while the outings gave them something to look forward to in an otherwise difficult life.

When I first joined the Society, it provided new footwear, socks and underclothing. Later on, we only gave out pairs of new shoes together with good second-hand clothing, which came from an occasional appeal or when a client passed away. I remember an ex-Guardsman who lived in a hostel in Central London, where they were all put out onto the street every day after breakfast. He turned up at Head Office once a year and he told me how he spent most days walking round the parks. So he wore out a pair of our shoes every year. His Regiment, I am sure, would have paid for the shoes but he never asked. A lady from SSAFA suggested that we should always take back the old shoes from a client as she was sure that some of them would continue to wear the old ones after selling the new pair in the nearest pub! I exercised my discretion in this matter.

All the Regiments and Corps in the British Army, including the now disbanded Irish Regiments, are contained in the Ministry of Defence list of charities. When a request is made, the Secretary of each Regiment will usually supply a general form of application used by all the ex-Service organisations. This is a neat way of summarising an applicant's details, his resources, Service record, and the reason for the request.

The Royal Naval Benevolent Trust acts in a manner similar to the Regiments in that an application can be made to them by individuals or by another visiting charity on behalf of anyone who has served in the Royal Navy or Royal Marines. The Women's Royal Naval Service has its own Benevolent Trust. In addition, if we consider all the Naval and Mercantile Marine charities there are many residential homes for old sailors, children's charities and other organisations which support naval dependants. All are listed in the Charities Digest.

For many years the Royal Air Force conveyed the impression that they were the best supported of all the Services when it came to benevolence. The Royal Air Force Benevolent Fund, founded in 1919, has always been comparatively well supported by the general public, especially each year when appeals are made commemorating the Battle of Britain in 1940. The aim is to care for past and present members of the RAF, of all ranks, and their dependants. The Fund has been prepared to make capital loans where income, in small sums, would be insufficient. Because of this level of support from the public, and the high profile sustained since 1940, they have been able to respond to the needs of ex-RAF personnel perhaps better than many of the other charities. Regular payments to relieve distress, block grants, and a generous attitude towards requests for small payments have been features of their response to clients in need.

I call to mind four elderly ladies, all ex-WAAF. One lady, whose medication caused her to be of a size requiring abnormal clothing, received a regular grant, as did a lady who had to burn expensive solid fuel in a draughty flat. One of the ladies was diabetic, and she received periodic funding towards the cost of special food. While the fourth lady was given an immediate payment when she broke her false teeth and suffering from considerable embarrassment!

A couple in Wales lived in a terrace house bequeathed to them by their parents. They had no capital funds and the roof leaked. Understandably the man of the house was in a state of anxiety. He would wake in the night, often in a state of panic, and would have to go round the house emptying strategically placed buckets full of rainwater. Builders' estimates were obtained and the Fund paid for all roof repairs on condition that the capital borrowed was repaid whenever the property was sold after the deaths of my client and his wife.

In another case, the RAF Benevolent Fund actually bought a house for one of my clients to live in and he paid them rent. This was because the local Council could only house this individual in a place that was unsuitable for his medical condition.

Mention must be made of Princess Marina House on the South Coast which takes both permanent and temporary residents and their carers.

Recently, however, there has been a noticeable exercise of stringency on the part of the Benevolent Fund, for reasons already touched upon. They have hinted that priority will be given to ex-Regular Servicemen and those from active Service situations as compared with the peacetime National Servicemen. They are also hesitating to consider refunding help initiated by another charity unless they are first consulted, and every alternative method of funding has been explored.

We have discussed at some length the availability of financial assistance for the disabled ex-Serviceman. The availability of care in ex-Service residential homes is probably woefully inadequate and admission depends on being in the right place at the right time when there is a vacancy. Costs may be shared between the charity owning the home and public funds.

Under the Community Care Act of 1990, from April 1993 every case has to have a care manager and the Local Authority must assess needs and co-ordinate Services from the Local Authority, and the voluntary and private charitable sectors. The NHS provides district nurses, community psychiatric nurses, health visitors and hospitals, while the Local Authority is responsible for funding placements in homes. Social Security benefits in these cases are no longer paid direct to the individual claimant. However, the basic aim is to keep as many people as possible living in the community for as long as possible.

There are four sources of help for the sick, disabled or elderly: the Local Authority Social Services Department, the NHS, the voluntary organisations (which include the ex-Service charities), and private homes and care Services which anyone can purchase. The Local Authority provides assessments upon request. For the elderly mentally ill there is a psycho-geriatrician in each Health District as a rule. The assessment entails looking at home circumstances, preferably using therapists and nurses. After this, a care package will be drawn up, probably involving a day centre, the CPN and possibly other organisations. A permanent long-term mental hospital is seldom recommended nowadays. Community Psychiatric Nurses, sometimes formerly employed on the wards at the large mental hospitals, are now based either in general hospitals or in health centres. However, they cannot be effective if social Services are inadequate. District nurses perform practical tasks at the home of their patients, although more and more is being done by Home Helps. District nurses are often overworked and sometimes, as with GPs, have little or no psychiatric training. As a further support, in some areas there might be an incontinence laundry Service.

Health Visitors offer health education for young families; they could certainly do more for the elderly. The Psychologist specialises in how the mind works, but he or she can help a patient who has lost confidence. And indeed some of my clients have been helped in this way. The occupational therapist can help by teaching life skills to the disabled in their homes. Meals on Wheels workers seldom have the time to see that the meals delivered are eaten properly. Many Social Workers prefer to work with the young and many times they have little in common with ex-Servicemen, especially if they are elderly. Day Centres provide social stimulation but some are obliged to refuse to admit the worst cases who exhibit unsocial behaviour, wandering off, or incontinence. Day Hospitals run by the NHS are similar to Local Authority Day Centres but they can carry out assessments, with access to consultants, and can take the more difficult cases.

Holidays are of no value to the patient who is confused but relatives, who understand, can help if they can take him for a break. However, it is often the carer who badly needs a break once or twice a year. The Crossroads Care Attendant Scheme, named after the former TV soap opera, is well worth knowing about.

Only 6 per cent of the elderly live where there are children.[3] There are dangers in having a sick grandparent in the family home; is there sufficient room in the house?, can they all get on with each other?, are there strains on the marriage? When I was eight years old, my grandmother came to live with us in the 1930s. She died when I was fifteen and there were problems, living in a small house and having to think twice before speaking one's mind in an argument. But in those days the idea of putting her into a home was totally unthinkable, and there was to be no NHS for another decade.

Today, there are several types of residential home. The ordinary old people's home, or registered care home, may be unable to accept a patient if the decision to admit him or her is left too late and the patient's worsening condition requires some nursing or supervision. If admitted earlier, once the person has become accustomed to the home, the patient's condition may remain stable and manageable for several years. Secondly, there are the psychiatric wards in a mental hospital or a geriatric ward in a general hospital. Thirdly, there is the private nursing home. Slowly, the old hospital wards are being phased out in favour of long-term specialist care homes for the Elderly Mentally Ill (EMI) or the Elderly Severely Mentally Ill (ESMI).

For the carer there are mixed feelings of relief and guilt when a patient is admitted into long-term care. There is also the anomaly of the long-stay hospital which does not make a charge on assets, such as the sale of the family home; whereas if the patient goes into a nursing home it can lead to the sale of the house unless the spouse is still living there. Numbers in this category are on the increase and there is substantial public disquiet and controversy as a result.

Local Authority homes were provided under Part III of the 1948 National Assistance Act. They replaced the old Poor Law institutions. Aneurin Bevan, Minister of Health in the 1945 Labour Government, dreamed of a situation where all old-age pensioners would live in decent communal hotels which offered care, dignity and companionship. In fact, the more active pensioners have tended to demonstrate their independence by staying in their own homes, and it has been mainly the unwell who have gone into Part III homes. It is said that over half the Part III residents have some form of dementia, although there is very little nursing care on the premises.

In 1990 the Government encouraged local authorities to purchase private care for patients. Some Local Authority homes became charitable trusts or were sold to the private sector. Privately registered care homes are not required to have any nursing staff but most of them do. Nursing homes must be registered with the District Health Authority and provide twenty-four hour a day nursing cover. A hospital can offer more by way of investigation, advanced treatment and rehabilitation therapy for the disabled. There are only a few establishments which are registered as mental nursing homes with twenty-four hour cover and a State Registered Mental Nurse in charge. The staff in care homes generally have only a little training in caring for the mentally ill.

A client with severe behavioural disturbance or persistent incontinence may not be accepted by an old people's home, in which case the only alternatives might be the long-stay psychiatric hospital (many are closing down) or a geriatric ward in a general hospital where the atmosphere may not always be cheerful or friendly. Old soldiers do not always enjoy the life in a hospital or a home where the majority of the patients are ladies with whom they might have little in common.

Sheltered flats run by the Local Authority may be suitable if the patient is near normal. But wardens are rarely trained nurses. The flats run by the Abbeyfield Society may be a more suitable alternative as they employ care staff on 24-hour cover.

Some proud people don't claim their allowances because it is felt to be accepting 'charity' and they believe the whole family has to be on the breadline before anything can be received from the State. Benefits offices are often gloomy places and their literature incomprehensible, especially if you are mentally ill. Recently, efforts have been made to improve both these defects in the system. Some clients refuse to spend any of their own money on obtaining help or let anyone else have access to it. This can be the result of feeling ill, insecure and vulnerable – all at the same time.[4]

The Social Fund is an uneven benefit depending on the amounts of money available locally. Outright grants have been replaced by loans which have to be repaid on a regular basis. The severely mentally ill, in certain circumstances, used to be exempt from paying Poll Tax whereas Council Tax, like the old Rates, is now based on property. One suspects that the mentally sick may have lost out here. The Disability Living Allowance may be awarded to those under 65; it covers mobility and personal care or supervision. Attendance Allowance is for anyone, young or old, severely disabled for at least six months; it is not taxable. There is an Independent Living Fund to help someone stay at home or, on return from hospital, meet the extra costs incurred. Finally, there is the Invalid Care Allowance for carers who devote more than 35 hours a week to their patient; it safeguards National Insurance contributions but the carer must earn less than £40 a week from elsewhere in order to qualify for this. In one case, a son was able to get this allowance for looking after his parents, both of whom were disabled and ex-Service. In another case, a disabled war pensioner was able to claim for looking after his severely disabled wife, on a full-time basis. These allowances for carers made all the difference to these two families.

Legal advice may be necessary to take charge of a confused person's affairs. A person of sound mind can give power of attorney to a relative or friend. However, a confused person cannot do this unless he gives enduring power of attorney before he deteriorates, and always provided that he understands what is happening. One of my clients used to frequently go down to the bank where he would ask to withdraw all his money, in cash! The Manager would then telephone his wife who would quickly appear and take him home. Fortunately, they had a joint account.

A relative may apply to the Court of Protection to administer a patient's affairs but this can be very expensive. The relative becomes "the receiver" who has to apply officially for everything; this is designed to protect the client from the scheming of a greedy opportunist. Alas, such persons exist, and occasionally I have had to deal with them. Informal authorisation can be made for small amounts of the sick person's assets through the DSS or a sympathetic bank manager.

The Mental Health Act makes it possible to have a person compulsorily detained or 'sectioned' who has no insight into his condition and is in danger. This can be temporary, for those who can recover, and compulsory admission for 28 days can be applied for by a social worker or a close relative. Two doctors are necessary – the patient's GP and a consultant psychiatrist; the period can be extended up to six months, and is sometimes extended voluntarily.

A Guardianship Order can be issued by the Local Authority to a close relative or to the social worker. This is less restrictive than a compulsory order and it gives no charge over finances. Section 47 of the National Assistance Act authorises the removal of the seriously ill or those living in insanitary squalor. A magistrate can issue the order, on the advice of the authority's medical adviser, but this is a power rarely used.

I have not yet mentioned the civilian charitable organisations for the mentally ill and the elderly. I have had clients who lived in homes owned by the Mental Aftercare Association, for adults recovering from mental illness. The homes are in the South of England and provide long-term care and rehabilitation for those sufficiently recovered to seek employment.

*   *   *

Ken was an ex-Serviceman who lived in a Mental Aftercare home whilst working for Remploy, a sheltered workshop organisation with branches all over the country. None of Ken's relatives would have him to live with them, although I always found him polite and well behaved. Perhaps I only ever saw the best side of his nature.

Ken served in the RAF where he suffered a nervous breakdown in India in 1944. He was medically discharged but never qualified for a War Disability Pension. Maybe Ken was one of those who should never have been accepted by the Forces in the first place. The Ministry of Defence have admitted that, in cases where there might be some mental weakness, it can be almost impossible to discover when a recruit is examined on first joining the Services. After the war he was in and out of mental hospitals, and he was looked after by his mother until she died. Ken lived quietly at the Mental Aftercare Home and his earnings at Remploy went towards paying for his keep. The balance was made up from money left to him when she died.

When Ken retired and qualified for a small Remploy pension and his OAP, he moved into a privately run home for the disabled and continued to visit Tyrwhitt House almost every year. When the stage was reached that Central Government no longer funded these visits, after April 1993, and the Local Authority alleged they could not afford to contribute (Government was supposed to have grant-aided local authorities to take over their former responsibilities but the money was never 'ring fenced' or specifically allocated to them), Ken managed to find a few hundred pounds each time from his savings, mainly from his late mother's money. This made it possible for the Society to allow him to come to Tyrwhitt House for a fortnight at a time even though the DSS no longer paid for it. Fortunately, the RAF Benevolent Fund always paid for his travelling expenses. Ken was a good example of a client who had benefitted from the Mental Aftercare Association and whom we were able to help by using a little of his own resources.

Another civilian mental health charity is the Mental Health Foundation which promotes, encourages and finances research into mental disorders of every kind. They promote community projects with care and after care for individuals and families suffering as a result of mental disorder or mental handicap. Thus the distinction is made between those born with some mental abnormality and those who become mentally ill in later life.

Another charity involved in mental health is MENCAP, the Royal Society for the Mentally Handicapped, which concentrates on children who were born mentally incurable. MIND is the National Association for Mental Health concerned with improving Services for the mentally ill and the mentally handicapped, and their families, and with the promotion of increased understanding of the nature and causes of mental illness.

I have encountered many clients who have benefitted from daytime clubs run by MIND, they have gained most from knowing that there are people locally who care. My greatest problem was always one of distance and whenever I left a client, we knew I might not see him again for six to twelve months, unless some dreadful crisis arose.

I have touched upon the great work of the Royal British Legion at Tidworth, Maidstone, and the Officers' Association, towards training and finding suitable employment for ex-Servicemen and women. The Regular Forces Employment Association has specialised in this work since 1885, the year in which SSAFA was founded.

After the First World War, the association went tri-Service and after 1945 they accepted applications from anyone with two or more years military Service. The high point of achievement came in 1948 when 80,856 cases were registered with the Association and of these 62,202 (or 77 per cent) were found work. As the Armed Forces declined in numbers, in an atmosphere of full employment over the next twenty years, only 8,500 men and women used the Service in 1979. Since then, the main task of the Association has been to encourage employers, especially those who are ex-Service themselves, to prefer to employ ex-Servicemen, and to go out into the world of work looking for openings for them. There are still branches in nearly every county, and a number of my clients have registered with them over the years. Interviews have been arranged between clients and the Association's officers. However, through no fault of the Association, only a few were ever found work. Not all of them turned up for interview, suffering from extreme nervousness, but this in itself should not cause our attempts to employ the mentally disabled to be abandoned.

One of the most successful sources of help for my clients, up to 1994, were two private nursing homes on the South Coast run by a qualified State Registered Mental Nurse who had served in the Royal Navy and worked as a nurse in the Society's residential homes. He thus had a rare experience of ex-Service personnel with mental problems. Some of my clients lived there for many years, and help was given towards obtaining or increasing a War Pension wherever possible. Some difficult disciplinary problems were handled with tact and firmness against a background of good living conditions. There was a sufficiently large number of ex-Servicemen to constitute a small community of around twelve; although a similar number of civilian patients also lived there. Attempts were made to attract younger ex-Servicemen and this was extremely useful when problems were encountered with veterans from the Falklands and the Gulf. A good working relationship with the Psychiatric Department of RNH Haslar, Gosport, was of enormous help when dealing with these patients.

For reasons not connected with the quality of his work, my colleague has recently ceased to be connected with disabled ex-Servicemen. Making the best use of all the expertise of the former Forces psychiatric nursing and medical personnel would be probably one of the best means of trying to resolve some of the problems raised in this book. I shall come back to this point later.

In spite of a comprehensive range of medical, nursing and welfare Services ostensibly on offer, my clients have consistently lost out, and this is one reason why their problems show no sign of a decline. The vision of a totally comprehensive NHS has dimmed in recent years, and attempts to save money on the Service, making it more efficient, have undoubtedly been detrimental to the mentally disabled. This has come at a time when all the ex-Service charities have been feeling the pinch – caused by increasing demands on them, falling revenues and reduced public donations. Care in the Community has not always delivered the most satisfactory solutions for my clients and, for obvious reasons, the community has not been the most suitable environment for some of the cases I have encountered.

Under recently implemented regulations, charities have been able to tender for the provision of welfare Services to a Local Authority. In this way, they receive much needed funds from guaranteed sources, on a regular basis, but they are subject to inspection and they must meet certain standards. There are fears that the charities may lose their independence and that, by using the charities' voluntary helpers, the Government is getting its responsibilities covered on the cheap. Daytime clubs for the disabled and some convalescent homes have recently been financed in this way.

I remember devoting many hours to completing complex application forms on behalf of the Society by offering short-stay convalescence in homes to local authority Social Services departments. I felt that the staff with whom I was dealing knew little about the problems of ex-Servicemen. I also felt that decisions often depended on how much money was left in the Authority's coffers after other commitments had been met. Thus the response to Care in the Community, as with many other local authority Services, was patchy. Once more, ex-Service people have some-times lost out when it comes to obtaining help.

In this chapter I have outlined the variety of sources of help to which the disabled ex-Service person may appeal. Unfortunately, clients of mine are suffering from a general shortage of resources trying to cope with an infinite demand from the public, set against a background of sometimes understandable prejudice against the mentally afflicted.

When it came to finding employment, I can well understand the problems of taking on someone whose temper may be on a short fuse or who may not turn up for work at all if he felt unwell, whilst superficially appearing to be perfectly normal, and therefore perhaps workshy and unreliable. My efforts to find work with the Royal British Legion were never a complete success although the Legion was undoubtedly a sympathetic, caring and under-standing employer of disabled ex-Service personnel. Those clients of mine who did manage to achieve a full life at work were either self-employed with a supportive wife who could take over on difficult days or who worked for another ex-Serviceman who perhaps shared, and certainly understood, my client's problems when it came to holding down a job.

Today's cut-throat business world is stacked against success-fully employing many of my clients. There is a general lack of job security and harsh competition for vacancies. The mention of psychoneurosis on a job application is often the end of the matter. Some who were employed, having omitted to mention their mental illness, were automatically dismissed after their disability was discovered. Because of this, some have refused to accept war disability pensions, or even apply for them. Many, on discharge, refused to admit to being ill, with detrimental consequences later on.

When it comes to the civilian medical and nursing Services, younger members of staff find it difficult to relate to elderly folk who were injured in circumstances they can scarcely compre-hend. In this respect, the ending of National Service in the 1960s has been detrimental to my patients. All ex-National Servicemen are now aged over fifty. I found it incredible when a Local Authority Social Worker stated that there was little or nothing wrong with one of my clients who was actually in receipt of a War Disability Pension for post-traumatic stress disorder following recent active Service.

The average general practitioner labours under an enormous work load, he or she is expected to be an expert on all things medical. Few are experienced in dealing with psychiatric cases; these will quickly be referred to a consultant. I have met cases where my client has been thought to have been malingering, when his physical condition was in fact being undermined by his mental state. Young relatives these days have little time for a crotchety grandparent; then there is the stigma attached to being related to someone who is thought to be mad.

The ex-Service charities have their own financial problems: declining numbers of legacies; increasing cases of hardship; and a Welfare State which, in theory, caters for everyone. Consequently some people ask why these charities need to exist at all.

The ex-Service convalescent homes and clubs are a sign of that unique fellowship which exists between ex-Service men and women of all ranks. Sadly, this fellowship cannot be enjoyed by some of my clients whose memories of the Forces are still vividly painful. Because of the risk of aggression or of other upsetting displays of emotion, many ex-Service homes cannot accept those who are mentally unstable; there is always the fear of upsetting the others and many homes do not have the nursing staff able to cope, or who are indeed qualified to do so. To provide permanent cover against the arrival of the mentally ill would require staffing levels which would increase residential charges for all the other patients. Most of the ex-Service homes ask the question, "Have you ever suffered from any form of mental illness?", and, sadly, they reserve the right to exclude some persons who have.

The ex-Services Mental Welfare Society's homes can manage cases of depression, mild schizophrenia and amnesia. But, even here, the line has to be drawn at cases of unsociable behaviour or perhaps requiring the constant care necessary for dealing with the severe aspects of, say, Alzheimer's Disease. Successful clients benefit from sharing their experiences. All have two things in common – their Forces Service and their mental problems. Thus generally they get on well and enjoy their stay, giving a period of respite for hard-pressed relatives and friends. Needless to say, these facilities are well used, and much appreciated, and the homes are always fully booked all the year round – which speaks for itself.

The next chapter will bring the various conflicts involving British Forces right up to date, together with some civilian situations which have produced comparable casualties in recent years.

Notes

[1] (1). The DSS, (2). War Pensions, (3). the Regiment, Corps or Service, and (4). the specialised ex-Service charities.

[2] Unless it can be shown that an application was made when the soldier was discharged and was not dealt with through no fault of the applicant.

[3] Elaine Murphy. *Dementia and Mental Illness in Older Patients* (Papermac, 1993). Much of what she says is very appropriate to the mentally disabled ex-Service person living with relatives or living alone in the community and not necessarily suffering from dementia – although some are. See also chapter 10.

[4] See the Age Concern booklet "Your Rights" and the "Disability Rights Handbook", published annually by the Disability Alliance. This is a clear, well-written book which I have consulted frequently. It is very important to always use the latest edition as some of the rules are changing all the time.

[5] References in the text to SSAFA (as opposed to SSAFA Forces Help) refer to the organisation prior to the ammalgamation with the Forces Help Society.

8

More Recent Conflicts: Post-Traumatic Stress Disorder and Civilian Tragedies

Most of my clients came from wars which lasted for a number of years. The Falklands Campaign and the Gulf War were over in a matter of weeks while Northern Ireland and Bosnia have seemed set to run as military operations for an indefinite period.

The Falkland Islands were invaded by Argentina in April 1982 after Britain had shown her intention of reducing our naval capability in the South Atlantic. When the invasion took place, Government House in Port Stanley was defended by only 79 Royal Marines although the planned reductions in the overall strength of the Royal Navy had not then been carried out. The surrender brought home to Britain a severe loss of national pride and 10,000 Argentinians occupied the islands. Mrs Thatcher was not best pleased with this state of affairs. The loss of HMS *Sheffield* and severe damage to HMS *Glasgow* showed clearly that, if the situation was to be retrieved, the islands would have to be invaded before the South Atlantic winter arrived.

A landing was therefore made at San Carlos Bay on 21 May but the overall strength of the British troops available could only just equal the strength of the Argentinians. This was after the arrival of the Second and Third Battalions, the Parachute Regiment, followed by the Second Battalion, the Scots Guards, the First Battalion the Welsh Guards, Royal Marines and Ghurka Soldiers. The troops who landed suffered severe attacks from the air and there was the loss of ships in San Carlos Water until air supremacy was achieved.

2 Para went south to Goose Green where 450 men defeated an enemy force which numbered 1,200. 3 Para made for Port Stanley and the Guards sailed round to Bluff Cove, where the *Sir Tristram* and the *Sir Galahad* were lost.

In all, we suffered casualties of 255 killed and 777 wounded. The Argentinians lost 1,000 dead and 11,400 taken prisoner. This was a notable British victory which bore evidence of superior military skills on the part of a highly trained professional Army. In future there was to be a permanent garrison of 4,000 on the Falklands, and a large airfield was constructed. The cost of the war was around two billion pounds. With the exception of Northern Ireland, future British military campaigns were to be as part of the forces of NATO or the United Nations.

The immediate effect of the Falklands War was that the public respected the Army more than hitherto, after notable feats of courage had been displayed. Good training had clearly paid off and the men had kept going in spite of casualties. The toughness of the Parachute Regiment was clearly shown and this image was to persist elsewhere and on arriving home. The speed of the Royal Marines' advance was applauded and the name for this, 'yomping', passed into the English Language. The Brigade of Guards were seen as far more than ceremonial soldiers.

The effect of mutilation and death on the soldiers varied from man to man. As will be seen, many psychiatric casualties only emerged years later. Some were to lose interest in the peace-time Army altogether; 'spit and polish' seemed irrelevant after some of the men felt they had seen everything, and they declined to sign on again. On reaching home some soldiers became more reflective and docile than they had been before; others became depressed and drank to excess; yet others became violent and displayed bouts of anger – as I was to discover. In terms of how we coped with our psychiatric casualties, the Falklands War was something of a turning point. For those of us at home, there were no air raids; we could only sit and watch some of the action on television, and it was all over in six weeks.

When I joined the Society as a Welfare Officer in 1984, they were interested, two years on, in ascertaining the consequences of the Falklands on the Society's case load. Hardly any cases had been notified. Two years later still, I only knew of two casualties. One of them required no attention, the other committed suicide. Many others were to appear later on over the next ten years.

*   *   *

Kevin and his wife were a very pleasant couple in their twenties. He had served for seven years as a Weapons Engineering Mechanical Ordnance Rating; he had been firing the guns on HMS *Glasgow*, three years earlier, when he saw the *Sheffield* go down with great loss of life. His own ship was severely damaged. He witnessed the effect of all this on his comrades at the time. Having left the Navy, Kevin was a full-time engineering student with hopes of a job with Plessey or Marconi. He had recently married. He asked me to call because he was experiencing nightmares, not only from the horrors he had seen, but also from wondering how many Argentinians he had killed when he fired the guns. Kevin's conscience troubled him as to how many young children would never see their fathers because of what he had done. His wife was a caring person but she felt useless because she had not actually shared the dangers with her husband. I said that if Kevin had not fired the guns somebody else would have. The Navy was bound to function in this way and did not rely totally upon any one person. I also told him that he would never know whether he had killed anyone; in fact he might have killed no one. I left various forms for them to complete and return to me, as Kevin was dashing off to college for his next lecture. He never responded and his wife, after Kevin had left, told me she thought they would try to ride out the storm of his troubles together. I wondered if they were successful. Reassurance from his loving wife might have done the trick. I hope it did.

I saw **Darren** with his Social Worker at a large mental hospital in the Midlands, just before he was discharged and declared fit for work. His problems had arisen as a result of a direct hit by an Exocet missile on HMS *Glamorgan*. We completed an application form for a War Disability Pension, but I was to never see him again.

Darren had joined the Navy in 1990 as an articifer apprentice. He was undergoing sea training in the computer room when his ship was ordered to the South Atlantic. He had a good education and would have become a qualified naval engineer in due course.

In June 1983, in a military hospital, Darren was diagnosed as schizophrenic, almost certainly because his mother had been so diagnosed whilst in a mental hospital. The condition was thought to be hereditary.

When I saw Darren, he described to me how he felt panicky, impatient and how he worried over little things. I suggested that we tried to get him some quiet clerical work to start with. Clearly, records submitted to Norcross for Darren's WP application would have referred to his schizophrenia and that of his mother, who was a widow with no other children. Mother was later found to be manic depressive, and not schizophrenic.

Before I could see him again Darren had received a letter of rejection regarding his War Pension. He took an overdose of his tablets and died as a result. On appeal, with a further consultant's report stating that Darren and his mother had, in fact, suffered from depression, not schizophrenia, we might have won his case. His mother was distraught with grief but felt comforted by the thought of a military funeral for a young man of 22 who had in fact died for his country. The local ex-Service organisations thought otherwise, declining to attend the funeral of a coward who had taken his own life. A friendly undertaker produced a Union Jack to drape over the coffin and a handful of neighbours went to the cemetery. For the War Pensions Officer and myself this was the first case of its kind we had experienced and Darren's death had a depressing effect on both of us.[1]

I first saw **Robin** in 1988 after he had benefitted from a six-week stay at Tyrwhitt House on the recommendation of a Welfare Officer from one of the Society's other areas. I visited him because he had moved to the South Coast to be nearer his estranged wife and two children. We had helped improve his pension from 50 per cent on discharge from the Royal Navy, to 80 per cent. He was about to start work as a technical housing inspector for the Local Authority. The Navy had been Robin's life since leaving grammar school at sixteen, on a technical apprenticeship. His ability was such that he became Fleet Chief Petty Officer for all the weapons control and guidance radar systems in the Task Force which went to the South Atlantic. From his base on HMS *Invincible*, the flagship aircraft carrier throughout the campaign, he was airlifted from one ship to another, sometimes after they had been wrecked by enemy action. Robin was acquainted with the officers and men on all the ships involved, and when news broke of casualties, he felt a keen sense of personal loss. Moreover, when Argentine bombs and missiles penetrated our defences, he felt personally responsible for these failures. Robin literally staggered from ship to ship and, when completely exhausted, often slept in the bunks formerly occupied by his dead comrades. This increased the strain on him considerably. He came through the Falklands War physically unscathed but mentally broken. After an inadequate period of home leave he was sent back to the South Atlantic sailing in ships of similar design to those lost or damaged, and as he moved around on board Robin imagined his dead friends were still there. His dead colleagues haunted him and the situation became intolerable. Robin was medically discharged with a Service pension and 50 per cent disability pension.

He fell out with his wife, had no peace of mind, drank heavily, and the Society took him on as a client. It was not long after taking the job for the Council, where he was his own boss, that drink got the better of him and he was dismissed. Robin just could not cope. Fortunately, he continued to receive psychiatric help from the military hospital and this was the best possible support he could have had. Although his pensions were substantial, there were money problems.

When I next saw Robin he was in lodgings. His wife and children were in a house but there was a substantial mortgage to be met. There were financial needs for the children, who were going into higher education. Robin earned a small amount teaching electronics to private pupils. At this stage, the South Atlantic Fund was approached; the mortgage on the house was paid off and a decent flat was found for Robin himself. I last saw him when we took part in a BBC television recording for Newsnight, dealing with the stresses of war. Robin displayed his interest in art, chess, music and mathematics. We discussed how his career had been ruined and the final camera shots were taken on the nearby promenade with Robin looking out over a sea he would never sail again. The programme was never broadcast because we were overtaken by more urgent items of news marking the start of the Gulf War.

I suppose Robin came to terms with his reduced circumstances but my opinion of this case was that there was a dreadful waste of great talent. More of it might have been saved if only he had been sensibly debriefed immediately after Robin returned from the Falklands for the first time. His wife was a capable, friendly person but she simply could not cope with his behaviour when he went through a bad patch. Consequently, more lives than just Robin's were blighted.

Jeff's case had some similarity with Robin's. He joined the Navy, aged sixteen and served all over the world on destroyers and frigates, eventually becoming a Petty Officer. He was on board HMS *Coventry* when she was hit and sunk off the Falklands. He came back to the UK and served for a further five years until he was posted to a ship which was identical to the *Coventry*. Jeff began to 'see again' all the places where his former comrades had been killed or wounded. It was discovered that Jeff developed PTSD immediately he set foot on that vessel. He received a 20 per cent War Pension after a year's treatment at Royal Naval Hospital, Haslar.

When I met Jeff in 1991 he was struggling to run his own business as a self-employed gardener. He lived with his wife and children in a smart new house and the South Atlantic Fund had granted them a capital sum to help reduce the mortgage to a reasonable amount. He had a modified Service pension and his wife worked, so they were not too badly off. Jeff still suffered from flashbacks and at times he struggled to keep on working. He felt nervous about asking his customers to pay him for the work he had done, but the idea of coming to Tyrwhitt House worried him in case he lost work as a result. Eventually, I reassured him that he could qualify for a treatment allowance, with expenses paid from the South Atlantic Fund. He had borrowed money from the bank to pay for his tools and the van. But, when he worried about this, he upset his family.

Jeff did eventually come to Tyrwhitt House where he benefitted greatly. His War Pension percentage was later increased. There were to be several more visits and he was able to abandon the idea of a regular self-employed occupation when his pension improved. This had the effect of calming him down and Jeff was able to manage reasonably well with occasional support from Haslar and the Society.

Huw was in the Welsh Guards and he suffered severe burns when *Sir Tristram* and *Sir Galahad* were set on fire. Physically, he lost the end joints of all his fingers and thumbs but was managing to use both his hands remarkably well in spite of their strange appearance. Mentally, however, Huw was in a bad shape. He never settled and I never succeeded in helping him beyond informing him how to contact me. He never did, and when I went to look for Huw he was always somewhere else.

I found that Huw had worked in a sheltered workshop before things began to go wrong with his marriage. When I saw him he was divorced and he became very upset at the mention of their young child. The South Atlantic Fund had helped but the outstanding mortgage on the family home still worried him. After showing some interest in Tyrwhitt House at first, he decided not to pursue the possibility of going there. Perhaps I should have been more persistent; he avoided me for the next three years.

His relatives told me how Huw drove around in his car, played golf and went fishing, but was clearly trying to avoid facing up to the realities of life. He was seldom in one place for long and spent his time staying with friends and relatives all over the country. He had received help from the South Atlantic Fund and refused to ask for any more. There had been relationships with females but nothing lasted.

Wayne was an NCO in a Royal Engineers airborne unit and although trained as a plumber, he was retrained as a specialist in explosives and bomb disposal. He was seriously injured in the Falklands when a mine blew off part of his leg. He was found to be 100 per cent disabled with multiple fractures and hearing loss. I found Wayne and his wife living in a comfortable home which the South Atlantic Fund had helped them purchase. The Army attempted to treat his injuries for no less than seven years before they discharged him.

When I saw Wayne he was still in great pain from where his leg had been amputated, and a further operation was under discussion as he was in a continuing state of distress. Wayne was able to do a sitting down job with a small electronics firm; he was also receiving psychiatric counselling from a military hospital. He and his wife had been offered a trip to the Falklands, but they were undecided as to whether to go. I suggested that it might do him good.

Shawn joined the Navy straight from school and served on *Invincible* as a steward during the whole of the Falklands campaign. When the *Sheffield* was hit, Shawn was flown on board to give first aid to casualties before the ship went down. He was later involved with large numbers of Argentinian casualties. He was discharged, 'Services no longer required', after he struck an officer whilst in detention. Shawn was treated for depression and then took an overdose of his medicine. It would appear that the authorities had no idea of what was wrong with him, so they let him go.

Shawn was unemployed for two years; with his bad Service record he was bound to have problems getting a job. Eventually, he obtained work in a warehouse, but he began to get headaches. Possibly because of them, he had a motorcycle accident; perhaps he should not have been riding one. When I met Shawn, he had short-term memory loss, some hearing loss, tunnel vision and paralysis of his left arm. When he was in hospital in 1991 Shawn had been in a coma for several weeks and it was discovered that he had suffered a subarachnoid haemorrhage. This was said to be an inherent condition of the brain and, because of this, he never applied for a War Pension. He was probably advised that he would never qualify for one. A further operation on the brain was carried out in 1993.

All the time, Shawn has had nightmares going back to the Falklands and, because of this, I attempted to have him seen by the Society's Chief Consultant with a view to initiating an application for a War Pension, suitably supported. I advised him that if it was felt that the brain haemorrhage had been inherent, his condition could well be judged to have been aggravated by Service in the Royal Navy. Haslar Royal Naval Hospital had first referred Shawn to me in 1993.

The Falklands campaign eventually led to an ex-member of the Scots Guards successfully suing the Ministry of Defence for £100,000 compensation in 1994. The campaign also led to some reflective writing which, with the general acceptance of PTSD as a recognised condition, was a turning point in Service attitudes towards some cases of mental illness.

Steven Hughes[2] was an Army medical officer in the Falklands in 1982. He returned to his civilian career in orthopaedic medicine as a consultant when he began to experience what he describes as a profound incapacitating panic attack with blind unreasoning fear. He thought he had accustomed himself to handling scenes of death at Goose Green and Wireless Ridge. Later he was in fear of losing his mind when taken to a psychiatric hospital where he resented not seeing the nursing staff in uniform, and suffered the indignity of having his tie and belt confiscated. Dr Hughes felt that the NHS system for dealing with mental illness was sapping what remained of his self-confidence.

Twelve days after discharge from hospital it happened again. This time Dr Hughes was sent home to recuperate, having successfully explained that the NHS psychiatric ward had previously done him no good at all but had probably harmed him. At this point he made contact with a naval psychiatrist who had also been with the Falklands Task Force. When he began to experience flashbacks to the Falklands it was realised that he had PTSD. At this point, his GP got him admitted to a Service hospital where the atmosphere was quite different in every way. Staff wore uniforms and patients were expected to return to normality, within or outside the Forces.

In order to do this, Steven Hughes had to come to terms with the memories of seeing his comrades and enemy personnel die, unable to save them after he had done his best. For him, Goose Green was the worst disaster; even *Sir Galahad* or Wireless Ridge were not so traumatic. Back home, he never seemed to have time to grieve for his friends. For him, treatment consisted in sharing, and thus dispersing, his mental injuries. Steven Hughes highlighted the dangers of PTSD in the caring professions. Dr Hughes is the co-author of a study of the incidence of PTSD in Falklands veterans five years on.[3]

*   *   *

This book is not a treatise in military psychiatry. And I am not professionally qualified in this field of knowledge. However, some of the discoveries made as a result of looking after clients were both revealing and of significance in better understanding the present-day problems associated with PTSD.

It is clear that PTSD, unrecognised as such, existed after traumatic events such as the Great Fire of London and the Plague, the American Civil War, the First World War, the London Blitz, and the Second World War. It has been brought on by civil disasters, violent crime and sexual abuse. In 1984 it was pointed out that problems reported after the Vietnam War arose following the stressful nature of combat in an anti-guerilla war together with doubts about the morality of some of the events that took place. There was also the popular unwillingness to welcome home those who had taken part; this heightened feelings of guilt already there.

The Falklands initially resulted in very few reports of PTSD, as my own experience showed. But by 1988 the popular press was complaining that the Ministry of Defence was ignoring the fate of the veterans. Reported casualties were small. The veterans were welcomed as heroes and most were back home within forty-eight hours of landing in the UK. The idea rapidly took hold that there had been very few psychiatric casualties.

The controlled study of veterans, five years on, enquired into the effects of: bad weather and action stations at sea, air raids at sea, damage to the ship, the wounding or death of a friend, being wounded oneself, assisting with casualties, disposing of bodies, directly killing someone, the receipt of an award for gallantry, being made out to be a hero afterwards, lack of understanding from friends and family, difficulty in talking to others about these things, strong emotional effects, loss of sleep, and problems in relating to colleagues and others nearby.

Half of those studied showed some symptoms of PTSD but only 28 per cent did not admit to any of them. About 85 per cent believed they had killed someone, 15 per cent had been wounded. Half of them reported an increased consumption of alcohol. All were likely to have an increased awareness of death from some quarter.

There were no mental health professionals with the land forces. Had there been any, the reported psychiatric casualty rate would have been higher than the initial 2 per cent; some might have been returned to duty with a decrease in the subsequent numbers of chronic psychiatric conditions. Twenty-two per cent of those studied showed full symptoms of PTSD five years on, while 43 per cent of Vietnam veterans were said to suffer from it after ten years (reported in 1982). This would indicate that there is a delayed onset of the condition, but American studies showed that the percentage does eventually reduce with the further passage of time. Some American studies of Vietnam related adjustment problems to drug abuse but there were, by 1991, no reports of drugs related problems with Falklands' veterans. The Vietnam veterans experienced an inability to talk with their families on returning home immediately after arriving in America; there was no period of adjustment or debriefing for them.

The Falklands situation was not unlike this except that the men were treated like heroes – a status most felt they did not deserve and could not expect to live up to. The Vietnam studies were based on those who complained of feeling unwell, whereas the British studies concerned soldiers who were still in the Army five years on, who had not been identified as disabled, but who had seen fighting Service on land in the Falklands. We did not know whether they did not see themselves as ill or simply decided not to complain because they regarded their symptoms as the normal outcome of what they had experienced; of if they felt that no one could help; or that it was not the done thing to complain. Many stayed in the Army because they needed the support of their friends with whom they had experienced traumatic events. Continued reminders of these events, for example in battle training, could have delayed the resolution of a distressed state of mind.

Of those who left the Services, some would admit to the illness, having reported its symptoms, while others would try to hide it. Military reminders would have gone except for TV programmes and anniversaries. Family support would be very helpful whereas lack of it could make matters worse. There could be perhaps a quarter of all Falklands veterans who are suffering from, being treated, or recovering from PTSD – perhaps 3,000 cases in all. The true numbers may never be known.

The Falklands was to lead to a milestone in legal history when a Corporal in the Scots Guards, injured on Mount Tumbledown, was awarded £100,000 by the Ministry of Defence in an out of court settlement, following a night-time battle against an elite enemy unit where nine were killed and forty of his comrades were wounded. He saved the life of a fellow soldier who was trying to shoot himself whilst swallowing blood. He also suffered personal injuries and, in 1990 he went berserk in Northern Ireland when he tried to kill his colleagues and himself. The Army, instead of counselling him and treating his mental condition, had him tried and convicted by Court Martial. The Corporal went to prison for two years, lost his pension, and was stripped of his rank. Military psychiatrists confirmed that he suffered from PTSD, his Commanding Officer wanted him back in the Battalion, but the case was lost.

The case was reopened in 1991 when it was claimed that the Army should have known that he had PTSD, and if reasonable steps had been taken his career would not have been ruined. The report and findings of O'Brien and Hughes, already published, had been ignored. Expert witnesses were ready to show that PTSD is capable of treatment, capable of diagnosis by asking a few straightforward questions, and that the condition prevents the sufferer who is generally unaware of his condition from leading a normal life.

Between 1982 and 1990 no one ever asked this man the questions relevant to PTSD, yet he had deliberately absented himself from training which bore resemblance to fighting in the Falklands. The Army had perhaps attempted to ignore these problems, hoping they would go away and not harm recruiting. The victim, after leaving the Army, was unemployed for long periods. He was divorced by his wife after years of drunkenness and violence, and he had no fixed abode. When the case was over, it was reported in 1994 that he was undergoing treatment in a psychiatric hospital. An MoD spokesman stated: 'The organisation is geared to treat trauma when it is encountered'. This was a little too late.

The *King George's Fund for Sailors Review* for April and July 1991 contained an article written by Captain David Hart Dyke RN, who lost his ship whilst escorting the amphibious vessels in San Carlos Water on 25 May 1982. They were hit by four 1,000 pound bombs which exploded deep down inside the ship. There was immediate flooding and fire, all power and communications were lost. Within about twenty minutes the ship was upside down horizontal before she sank. It was remarkable that, but for the nineteen men tragically killed by the bomb blast, 280 got out alive which Captain Hart Dyke attributes to their good training, morale and discipline. Men climbed through the heat and the smoke and calmly abandoned ship; as the vessel was about to roll over, the Captain saw everyone else off the wreck, walked down the ship's side, jumped a couple of feet into the water, and swam to a liferaft.

The shock of this had a profound effect on him. On the way home, the ship's crew kept together; only they knew how the others had suffered and they benefitted from shared experiences. They discovered it was wrong to suffer in silence and better to relate their experiences to other people. The Captain found it awkward, back home, to find others who perhaps trying to be considerate, avoided mentioning the Falklands. It took him two years to feel that he had recovered from losing his ship, without ever forgetting those events. When such happenings are suddenly over, it is hard to adjust to a new environment where there is no war and no need for decisive leadership. These longer-term effects need to be more fully understood.

Those who fought in the Second World War commented that they had little option but to recover from one scene of battle by taking part in yet another. However, the effectiveness of the weapons is much greater now than it was fifty years ago. In spite of all the accumulated experience of fighting since 1939, it was pointed out that technical training in the Navy over the past fifty years has been excellent in content, but no one, not even those decorated for gallantry, ever spoke about what active Service was really like. Perhaps it was thought to be unmanly to speak of stress trauma and how to cope with it. Training, to be realistic, should include some of this.

In spite of the feelings of Captain Hart Dyke, the 'stiff upper lip' did have some support. Horror was felt at the idea of sending out a naval psychiatrist to counsel the senior crew members of a ship involved in an incident in the Gulf War. Perhaps this was necessary because modern fighting men are not as tough as they used to be. Perhaps every generation will point to the one before saying 'They don't fight like their fathers did!' Present-day talk of men crying to relieve stress is still regarded as not the done thing.

Only nowadays are disasters met with cries for sympathy and counselling. As I have said, war is much more frightening now than it was; the media are more involved, and more is known about the effects of trauma and the dangers of not treating it.

In the First World War, "they shot them, didn't they?" Battle of Britain pilots laughed and joked about their exploits, but the effect on some of them was to be cumulative. Perhaps we complain today because there are more people who will listen. It was no use complaining when you were in the 'Forgotten Army' in Burma for years on end. Again, the cumulative effect of these conditions did take its toll on the men. It is arguable that modern-day, western-style living makes men soft physically and mentally. But the Special Forces would not agree. Life today is comparatively easy and leisure orientated, but it is materialistic and spiritual values count for less than they did so there is less to fall back on than used to be the case. Moreover, it used not to be the done thing to admit to any mental disorder. Are present-day people spineless on the whole? Do they fight like their fathers did and get on with things without complaining? Psychiatry has, and is going to have, an increasingly important role in modern warfare.

Over the past thirty years Northern Ireland has demonstrated the need to withstand an unseen urban guerilla enemy, far removed from set-piece battles against a recognisable foe. Casualties have come from there in a steady stream during the whole of this period. Northern Ireland veterans have cracked under the combined effect of more than one type of military stress. British troops in Northern Ireland have had to separate the two communities by peace lines and standing, as it were, in the middle, they have been welcomed and hated at different times by both sides. At the same time, two wings of the IRA have attempted to intimidate the civilian population while Loyalist para-military forces have carried out attacks on the Catholic community. Two achievements have enabled the Services to cope with all of this. The Army has created its own fortified areas into which the men could retreat and live in a sheltered environment before embarking on the next dangerous assignment. There was at first a lack of co-operation with the RUC, and military intelligence was poor at the outset. The security forces gradually learned to pull together, first through bomb disposal and secondly through military police operations, sometimes under cover. The imposition of direct rule from Westminster angered the Protestants but gradually security screening did pay off; and when the IRA went to South Armagh, the SAS followed. Up to 11,000 troops at a time have been deployed there.

The serving soldiers are frustrated when there is no visible enemy; an attack can occur anywhere and the assailants can simply melt into the background. Soldiers are annoyed by the Rules of Engagement under which, unless they are actually under attack, any casualties inflicted will be treated as civil crimes. No one has ever actually declared war. The Province is a part of the UK with all the familiar road signs, British public utilities and civilian uniforms. Battle training at home has made the men impatient to get into action; in Northern Ireland, this is the one thing they must learn not to do on impulse. Service in the province has been a source of PTSD for almost thirty years. It is a direct result of apparent normality, followed by sudden violence, followed by hatred. The men cannot relax openly and it is this 'knife edge' existence which leads to stress disorders which can appear at any time, perhaps years afterwards. It has been said that periodic debriefings could help normalise emotions.[4] About ten men per unit might have persistent mental problems requiring psychiatric treatment, perhaps 140 a year, or several thousand over the past three decades. Many soldiers are very reluctant to admit to any of this for fear of letting down the Regiment or blighting their careers in the Army. This is another aspect of the 'stiff upper lip'.

Below, I indicate the types of casualties with whom I came into contact or attempted to help in some way. On 27 August 1979 two bombs in succession killed sixteen members of 2 Para at Warrenpoint as they travelled in the back of a four-ton truck. One man who later went to the Falklands claimed that this incident never affected him emotionally, but we do not know whether the cumulative effect of all of this took its toll.

*   *   *

Keith was a senior NCO in the Parachute Regiment. I saw him in the Psychiatric Department of a major civilian hospital, at the request of the Senior Consultant. Keith had completed his 22-year regular engagement, serving seven times in Northern Ireland, and in the Falklands. He had been in the incident at Warrenpoint. When I met him the hospital clearly did not know what to do with him next as his behaviour was near normal after six months' treatment as an in-patient. His wife had left him because of unreasonable behaviour and, on leaving the Army, he went into security work. Then he became a lorry driver, but he assaulted two of his workmates. The firm had to dismiss him immediately but agreed not to take legal proceedings against Keith provided he took psychiatric treatment, which he did. He had lost his flat. He told me he was comfortably well off on his full Army pension and he had no intention of being branded as insane by applying for a War Pension. Without such a pension there was no way the Society, under recent rules, following the Community Care Act of 1990, would admit him to Tyrwhitt House free of charge. Keith said there were more deserving cases than his. There had been several outbursts of violence but I suspected that medication was keeping him calm, provided he continued to take it.

I therefore took steps to have him admitted to Haslar where he would receive treatment by way of a military psychiatric debriefing to help him come to terms with his problems. I never saw him again.

Whereas Keith was involved in picking up the bodies of his friends at Warrenpoint, **Vic** was actually on the truck when it blew up and, because his comrades were killed while he was blown clear by the blast, he felt guilty. I never saw Vic but I answered a distress call from his wife who told me her husband was depressed and could be violent. He had set up a security business, installing burglar alarms, when he left the Army. He took out a £10,000 bank overdraft to set it all up. However, Vic went berserk, assaulted a customer, and was sent to prison for a year. All along, he refused to admit he was ill. He would not ask the Regiment for help, nor would he apply for a War Pension. I wonder how many more there are like Vic.

Vic refused to see me and his wife said he was living in a hostel in Central London, trying to run away from his creditors. I therefore took what steps I could, through her, to get him admitted to Queen Elizabeth Military Hospital, Woolwich. She never contacted me again.

Bob came from a military family and, after following his father's postings, as a child, he too joined the Army as a boy soldier. He received a lump-sum payment when he took his discharge after twelve years' Service. Bob always refused promotion. He tried a number of occupations but never settled to anything. When he applied for a War Pension for hearing loss, his War Pensions Welfare Officer referred his mental condition to me. This was when behavioural problems began to emerge. He was greatly distressed by an incident in Northern Ireland when a child was shot by accident and some of his comrades were killed. I suspect he thinks he killed the child when the soldiers returned fire. Bob blames himself for this and for having survived. He was having panic attacks and nightmares, and was living in a state of fear.

Bob's wife had responded to some local publicity regarding a War Pensions Open Day when members of the public were invited to meet the welfare officers and discuss their problems. They were very short of money, relying on Social Security payments to bring up two teenage boys. They were able to receive support from the Society towards the addition of PTSD to his War Pension and the family benefitted. Bob himself eventually felt that he had received some recognition for his suffering and this helped him adjust to it.

Ernest was a highly skilled Ammunition Examiner in the Royal Army Ordnance Corps. I had served with him in the Far East where he was a Staff Sergeant and was highly regarded by all who knew him. He eventually served for thirty-one years but his final posting to Northern Ireland, working on bomb disposal, ruined his life.

When he came out, aged fifty, he could not settle down, and the traumatic experiences working with bombs and high explosives began to affect him in spite of his long Service working in these dangerous jobs. Ernest was full of ideas. He tried running a bakery, having a stall on the market, then he became a caretaker – but it all came to nothing. He felt ashamed of himself, drank heavily, which was completely out of character, and he left his wife. I saw him in Cornwall where he had plans for opening a guest house. Again his plans collapsed. I followed him to Sussex but this time he avoided me and I never saw him again. He might have benefitted from what the Society had to offer, but I suspect that Ernest was afraid of letting me down.

The Northern Ireland scene was particularly traumatic for people who actually lived there. And **Billy Douglas**, who joined the Ulster Defence Regiment, was no exception. I met him in London where he lived with a very caring lady who has stood by him ever since. He was fond of regimental life and his flat was adorned with military badges and other souvenirs, all highly polished. He was Guard Commander when his unit faced a direct attack from terrorists; he saw his comrades lying wounded and dying. On his return from hospital for pneumonia he found his home in Belfast had been vandalised by terrorists.

He and his family simply had to leave the Province as soon as Billy was discharged, but that was only the start of further problems. He developed agoraphobia and, when his wife went out to

Billy Douglas, then and now.

work, he would sit in the house all day, afraid to go out in case the bombers were after him. He was living on his nerves. Billy did pull himself together and got a job as a gardener for the local Council. I suggested we might try for a War Pension. At this point, his wife fell ill and had to give up work. They are a very friendly couple who deserve our support but who are content to live quietly, which is understandable after all they have gone through.

Ryan's case took many years to resolve as his pensions application in respect of psychoneurosis and hypertension had been refused in 1988 on the advice of a consultant who did not have all the facts in his possession. I entered the scene in 1992, at the request of the local Community Psychiatric Nurse.

Ryan joined the Army with his brother, who later became a sergeant major in the same regiment. He did a number of tours in Northern Ireland, also a short period in Belize. During his final posting to the Province, he was in and out of hospital with uncontrollable blood pressure and he opted to leave the Service, while his brother stayed in. He worked in a factory for five years but, following the death of his mother, he could not cope with his feelings and tried to commit suicide. His wife divorced him. He described to me how he tried to switch off his emotions, but by then he was hearing voices in his mind. It was at this time that he was refused a War Pension.

I found Ryan liable to mood swings, hesitant in speech, and having to force himself to go out of the house. His sleep was constantly interrupted by nightmares. We took him into Tyrwhitt House where he had a successful stay; he was invited to extend his visit from two to four weeks. I helped him compile a diary of events in Northern Ireland which had upset him, including an incident when he was cut off from his patrol, alone and cornered by enemy fire. This evidence was shown to the Consultant Psychiatrist at a nearby hospital.

I concluded that Ryan had PTSD which had been aggravated every time he had another shock of some kind; for example, when he injured his back in an accident whilst working in the factory. He received compensation for this but the trauma was never recognised. We then asked Ryan's brother to give a written account of what he knew of Ryan's experiences in Northern Ireland. The civilian consultant and the Society's consultant were both able to support the case for Ryan's application to be reconsidered; this time we were successful. I doubt whether Ryan will ever work again but he enjoys periodic stays at Tyrwhitt House. Financially he is better off, and recently he has remarried. His new wife is a very caring lady. I was pleased to be able to help in this case but my involvement took place over a number of years. It all showed that patience and perseverance can succeed and this was my reward.

Barry was in the Parachute Regiment in the 1970s. He was injured by gunshot wounds in Londonderry. Barry had a 50 per cent pension for physical injuries and, when I met him, he was a voluntary helper at a local home for the elderly. He became known to SSAFA who referred him to us because he had mental anxieties. In addition to his wounds, which severely restricted the use of his right hand, Barry was afraid of crowds and afraid of being followed. He could not stand queues of people, and he suffered from nightmares.

The Army had kept him in for four years of treatment and rehabilitation before he was finally discharged; he was due for further treatment on his hand. Barry had tried to work until he had a complete mental breakdown in 1992, and had not worked since.

An application for the addition of PTSD to his accepted disabilities was lodged and, on the strength of this, anticipating that he would be successful, he was admitted to Tyrwhitt House where further information in support of his case was obtained. Barry and his wife and child were living with relatives, and were anxious to obtain a place of their own. With the support of their GP, they were on the waiting list for a council house. The local psychiatric Service was unable to help at that time, but with the Society's continuing support it was possible that Barry might return to work, have his pension increased, and move to more suitable accommodation with his family.

* * *

One of the features of the Gulf War was the small number of actual British casualties. It was therefore easy to assume that the campaign presented few difficulties and few problems afterwards. Neither of these assumptions was to be correct. It was originally expected that, based on 30,000 British Forces in the Gulf, there would be 1,500 casualties of which a quarter would be killed. In fact there were over 45,000 British troops by the end of the campaign and there were 49 Field Surgical Teams employing 200 consultants. Medical reservists were called up and some friendly countries also contributed surgical teams. All casualties were to be returned to the UK. The dangers of chemical warfare was to cause complications, some of them far-reaching. An attempt was made to warn the press to maintain silence over the extent of casualties.

The Allies faced an Iraqi army of half a million men and there was much public sympathy for our men at Christmas 1990. It was discovered that some of the vehicles were unsuitable for desert fighting, and there was a political crisis at home in the November. One of John Major's first acts was to pay a visit to the Gulf and take steps to reinforce the UK Forces there. The Americans had 500,000 men out there, over half their total armed forces. Conditions were not ideal for our troops, who encountered boredom and loneliness with eleven hours of darkness in blacked-out tents in isolated desert conditions. At first there were only occasional film shows to relieve the boredom. At least one man cracked up under the strain and wandered off into the desert. But eventually he gave himself up.

The Authorities did try to improve conditions and attempts were made to boost regimental morale whenever new personnel arrived to make the numbers up. A routine of training and the use of a relaxation camp helped considerably. General de la Billiere felt that the ban on alcohol, in deference to Saudi Arabian tradition, was a good thing; anyone caught in possession of it was sent home in disgrace. The Forces Broadcasting Service started on 16 December and free parcels began to arrive at the same time. Then there were telephone cards and a visit from the Prince of Wales.

Anti-biological immunisation made some soldiers ill. Wearing NBC suits (Nuclear, Biological, Chemical warfare) heightened tension. Tablets designed to reduce the effects of exposure to nerve gas gave some soldiers diarrhoea. When scud missiles fell on Ryadh, everyone became apprehensive. The Government policy of reducing the strength of the Armed Forces, 'Options for Change', meant that some soldiers were to be declared redundant when the campaign was over. Those who knew of this were not likely to be pleased at the thought of their careers being cut short at a time of unemployment at home.

Some military operations began before the main offensive. The SAS were used to cut roads, create diversions behind enemy lines and take out scud missile launchers. Low-flying attacks facing severe 'triple A' (Anti-Aircraft Attack) were made on Iraqi airfields and there was a continuous anti-aircraft watch kept on board naval vessels for weeks on end. All of these situations led to a cumulative strain. Then there were power breakdowns on Challenger tanks. There were morale difficulties at the Battle Casualty Replacement Base where personnel of all skills and trades were billeted together with nothing to do but wait for casualties to occur. Four SAS men lost their lives countering the scud missile raids and they earned the highest praise from General Swarzkopf, the Supreme Allied Commander. The Air Force and the Navy had been fighting for some weeks before the ground attack began at 4 a.m. on 24 February 1991.

After ten hours of fighting, ten thousand prisoners had been captured. There was tension inside the armoured vehicles where very young men were frightened in spite of months of training. There was always the fear of poison gas being used; then there was darkness and heavy rain. For some, it was terrifying being inside a tank at night, firing at targets only indicated on thermal sights. The rapid forward movement ended in two Warrior vehicles being accidentally destroyed by American aircraft fire. They blew up in two balls of flames. The advance went much further and faster than was ever imagined, but after three days and nights without sleep, levels of reaction were much reduced. The Fourth Brigade had been in contact with the enemy for 54 hours and had travelled over 350 kilometres. The REME had completely changed 74 Challenger tank power units, a tremendous achievement.

In the end, the British division took over 7,000 prisoners; it took three battalions to guard them. There were scenes of mass destruction of Iraqi soldiers, as a result of the prior bombing, and bodies were strewn everywhere. The prisoners were half starved, and infested with lice. Their wounds were untreated and they were convinced they would be tortured. There were many individual acts of bravery under fire and in the minefields. Over 60,000 prisoners were finally taken by the Allies. In spite of extensive TV coverage, few people in Britain had any idea of what fighting this war was really like.

The power and range of modern warfare makes it more stressful than the one-to-one combat of earlier generations, especially when those taking part have very little idea of what is going on. This applied particularly to the RAF ground crews for most of the time.

Training for both psychological and physical warfare may well be needed in future. There is still the problem of military leaders who are of the opinion that being affected by the stresses of war is an unacceptable sign of weakness. This, in time, can cause even greater stress amongst those likely to suffer in this way; there were some Iraqi troops so traumatised that they were unable to fire a shot, terrified by the weaponry and by their own heartless officers. Today's small numbers of highly-trained regular soldiers find it hard to meet anyone to whom they can relate their problems, and who is not in the Army. I now know why so many of my clients were so pleased to see me, and how it was that some of them never stopped talking all the time I was with them.

The RAF Tornado attacks consisted of two and a half minutes of complete panic over the target, surrounded by four and a half hours of comparatively calm flying and preparation. There were plenty of stressful situations bottled up for later on.

Popular attention has been shifted from all this by the discovery of Gulf War syndrome, which even appears to affect the Servicemen's children physically. The mental effects of this may be only peripheral. By the end of 1993, only one client from the Gulf was known to me and the situation was not unlike the aftermath of the Falklands ten years earlier.

The debriefing of the aircrews must have helped but ground staff seldom knew what was happening. Then there were fears of terrorist attacks from the PLO, a situation not unlike Northern Ireland. Any RAF personnel missing were automatically presumed dead by their comrades. In the Air Force, in spite of Falklands experience, there was no preparation for battle shock casualties; when operations began the younger, inexperienced men were weeded out from their squadrons to be replaced by more experienced strangers. And so, another Falklands lesson was ignored, that of keeping men together which helped preserve morale. Men soon felt irritable towards colleagues whom they did not know well.

The fear of chemical warfare was combatted by the issue of NBC suits and the learning of attack drills. There were good medical facilities, but most were never needed. The mass evacuation of casualties to hospitals in the UK never took place. The Falklands and the Gulf produced some macabre humour towards the enemy dead which may have relieved tension at the time but, with some photographs taken, may have revived bad memories later. The actual number of Iraqi dead was never published; it was impossible to count them all. This whole experience was to have an effect on Falklands and Vietnam veterans who read about it and watched the news on television.

PTSD in the Gulf was generated by the waiting and the threat of chemical warfare. Another "trauma" was knowing that there were teams whose job it was to collect the dead bodies. Battle Recovery Units (BRUs) were the first field psychiatric units to be deployed in action since the First World War. They had few patients, but there was a twenty-four hour consultation Service available – a new and progressive aspect of modern warfare.

Later on, the idea of offering treatment for PTSD was forgotten as commanders saw little need for this, having won the war. The field psychiatrists from the RAMC were not heeded; the men went home on leave and units were broken up. The men were frustrated at not finishing off Saddam Hussein. Thus there was a general feeling of anti-climax, relief at having survived, and memories of horrors inflicted on the enemy. All this added up to an emotional cocktail, perhaps a time bomb. In 1992 there were officially 500 known cases of PTSD and the psychiatrists said this number would increase. Those who had any mention of this on their record were not promoted, in future, as a rule; many of them left the Services anyway. The Navy did send teams of psychiatrists to the ships in the Gulf and the RAF treated the British hostages coming out of Beirut, as well as those aircrew who had been captured by the Iraqis. Northern Ireland was producing about 140 PTSD cases a year requiring formal treatment.

The treatment of our returned prisoners after the Gulf War was a turning point in the history of military psychiatry. It was clearly demonstrated that individuals who had been imprisoned and even tortured could be returned to operational duties whithin a short space of time, provided that they went through a 're-entry process' after being released.

The work undertaken at RAF Lyneham and RAF Wroughton by military psychiatrists following the repatriation of the hostages from Lebanon (John McCarthy, Jackie Mann and Terry Waite) fully demonstrates the success of what was achieved.

Hugh McManners, in the final chapter of his book *The Scars of War*, hoped that a Combined Services PTSD Centre might have emerged from the UK's war experience. However, it appears to have been thought that educating the men about mental stress would lead to a generation of cowardly weaklings. Attitudes such as this went right back to the First World War. Nonetheless, we know that psychological debriefing after a traumatic event can help enormously. One suspects that the Ministry of Defence was aware of a possible increase in the number of claims for compensation, therefore no accepted centres of treatment for PTSD were ever set up.

On the other hand, in recent years, the War Pensions Agency has been able to embark on a number of measures to encourage veterans to apply for pensions by increasing their awareness of what is available to them. As a result, the overall numbers of war disablement pensioners has increased substantially in the last few years.

*   *   *

I met **Thomas** towards the end of 1993, shortly before I retired. His father had been a regular soldier and it was he who wrote to me. Thomas had been out of the Army for two years, after serving in the Gulf, where he was eventually employed in collecting enemy dead bodies. He had never told anyone about this and he was glad to tell me some of it.

Since leaving the Army, Thomas moved quickly from one job to another, as a builder's labourer. But he was unable to settle. I put him in touch with the Regular Forces Employment Association and informed him about his rights regarding a War Pensions application, and how to proceed. Finally, I told him about Tyrwhitt House, not expecting him to ask to stay there in the immediate future.

*   *   *

Gulf War Syndrome has affected some men and women who served there, as well as their young children who were born after the war. The problems experienced are mainly physical problems, as opposed to the mental. However, the combined effect of the Syndrome and the mental stresses producing PTSD could possibly interact. This could be the case if the British authorities do nothing to recognise those who are now claiming that Service in the Gulf has affected them medically. At the time of writing, an official enquiry is being set up to study the evidence submitted from the victims.

Early in 1994, it was announced that there were 250 claims for compensation being lodged for abnormal symptoms being experienced by some Gulf veterans. The MoD at the time refused to recognise any connection with Service in the Gulf. A typical sufferer found that he was completely exhausted after only three days' work in an average week, and he had severe body pains, diarrhoea, and an aversion to noise. Service medical officers said that these ex-soldiers must have contracted a virus.

Two years later, the number had grown to 700. All complained of memory loss, fatigue and swellings; these were alleged to be connected with injections and other medication which had been administered to counteract the effects of chemical warfare. The Americans reported greater numbers of personnel affected in this way, and they proposed to carry out a major investigation.

There was evidence of defects in children born to personnel who had served in the Gulf; possibly as a result of side-effects of the cocktail of injections which had been administered. There had been a lack of informed consent to what they had been given. A number of babies had died, or were miscarried, or were born with horrendous defects. The mental strain on the parents is tremendous and there are cases where the veterans themselves have become depressed, largely because their physical problems are not understood. Nor are they yet recognised. This is a situation which will not be resolved until the authorities on both sides of the Atlantic are able to make informed decisions as how best to cope with what has happened since operation Desert Storm.

Numerically, the British Army's involvement in Bosnia has been small by comparison with campaigns in the Falklands, the Gulf, or Northern Ireland; yet the likelihood of PTSD cases arising is undoubtedly there. The situation has had similarities with Northern Ireland in that we have been attempting to keep the peace between intractable enemies. The soldiers are not permitted to go onto the offensive; they have to report incidents without becoming involved. The United Nations Forces, of which the British were a part, were always outnumbered, out gunned, and coping with a hostile terrain in bad weather conditions. It was difficult to maintain our own supplies and communications, and there was always the danger of sniper fire. Then there were the refugees attempting to escape from hostile forces in places where all had formerly lived together in apparent peace for decades; this was until the collapse of the Communist State of Yugoslavia and the ending of the Cold War. We were, in effect, attempting to pick up the pieces left by a state of undeclared civil war. The British Army could go back in time to the partition of India in 1947 and the parallels were there, except that then, we knew that we would be withdrawing permanently in the forseeable future.

In Bosnia there has been no such outcome although the arrival of large numbers of American Forces and the military functioning of NATO has recently made conditions safer for our troops. In spite of recent peace treaties, basic feelings of hostility between Serbs, Croats and Moslems still remain largely unresolved. Troops serving with the UN Peacekeeping Force were on a six-month tour of duty which included a 17 days period of rest and relaxation for all ranks. This was essential in view of the sniper-ridden conditions. Attacks on our vehicles were not uncommon; mountain roads were dangerous, could be unsafe in winter, and there was always a sheer drop on one side. The death of a driver, killed by small-arms fire, made the others feel guilty. His body was evacuated from his vehicle while it was still under fire.

At one time, five coaches of refugees disembarked as they fled from the Serbs at a crossing point. They were old women and children, and the soldiers had to carry the little ones over the front line minefields. By nightfall, any refugees not on the transport were shot at until British Warrior tanks returned the fire. All this took place while children were lying in the road, attempting to take cover. In another situation, our armoured vehicles were completely surrounded by refugees as the Bosnian Serbs shelled them. The troops found this very upsetting. Soldiers carried the wounded to shelter, but a baby's head was blown off by shrapnel. Other children had their legs severed. The British Medical Officer did what he could while the people, understandably, panicked. It is not surprising that Lt-Col Bob Stewart's book is entitled *Broken Lives*. Then there was the lady interpreter, well known to our men, who was deliberately killed, shot in the head by a sniper. The troops found this very upsetting. Innocent people were killed outside their homes, and when a truck full of ammunition was blown up near Vitez, hundreds were homeless, many more were slaughtered.

At Ahmici a complete Moslem village was wiped out; houses were fired, people and animals massacred. When UK forces arrived there, the soldiers had to examine every building, looking for survivors. The men were choked with emotion. A group of 150, about to be marched to their deaths, was intercepted by British soldiers. A child of six described the death of her parents outside their house. In all these situations the men of the Cheshire Regiment opened fire only seven times. They killed four men. Although British casualties were few, such memories are bound to live on in the men's minds. All this was similar to those situations when British soldiers entered the German concentration camps in 1945.

Britain played no part in the fighting in Vietnam, but the mass destruction of this war with large numbers of civilian and military casualties was to lead to PTSD being recognised as a likely consequence of war. As a result, many who had survived similar situations during and after the Second World War were to be diagnosed as having this disorder. Many such conclusions came years after repatriation to America and Australia; regular and conscript soldiers were found to be equally liable to suffer from a poor work record, violent outbursts and alcohol abuse.

*   *   *

One such case was referred to me during my visits. **Colin** sought the help of the Society. He was an ex-soldier who suffered nightmares. He had been treated at the local psychiatric hospital where he had been supplied with medication; in view of his unusual military record, there had been no approach to any of the ex-Service organisations. I found Colin very friendly, physically fit and, at 60, a batchelor trying to obtain work as a labourer. He had always been single. He had no family in this country and had always tried to work all his life.

Colin, who was born in England, was called up at the age of eighteen and served in Germany. He signed on, went to Malaya, and then to Northern Ireland. In 1960 he joined the Australian Army for six years. He was in Papua-New Guinea, and then did fifteen months in Vietnam with the 82nd Airborne Regiment. He only hinted at the terrible events he had witnessed and taken part in, but he told me he had been decorated for Gallantry in the Field. After all this, he rejoined the British Army for a further period, but peacetime soldiering was, understandably, by this time tame and unrewarding.

Colin worked in Algeria, and then Rhodesia, coming home for periods of work in England. Not surprisingly, when I met Colin he was being treated for depression. This was only part of his mental problems. At this time professional studies were being published on the effects of the Vietnam war on Australian Army personnel.

I did two things; I put his name down for Tyrwhitt House and I contacted the Military Attaché at Australia House who was sympathetic to the idea of applying for an Australian War Disability Pension. It was fortunate that I had, at one time, served with the Australian Forces in Malaya. I accordingly wrote to their Department of Veterans' Affairs, in Tasmania. Medical reports were submitted from our British consultants at the local hospital and from the Society. A one hundred per cent total disability pension was granted with very little fuss at all. The complete total of all fees and expenses incurred at Tyrwhitt House were reimbursed, and there was a lump sum in back pay. Colin has settled down in a Local Authority flat, with a warden on hand. He now has no financial worries. He does voluntary work and our helping him has been totally successful.

*   *   *

We have touched upon those who have been traumatised as a result of dealing with unexploded bombs. Another type of potentially hazardous military Service, which is never publicised because of its very nature, is the gathering of intelligence. It may be directly connected with military operations in progress but always concerns events and organisations likely to threaten the security of the State.

Anthony was a brilliant linguist who joined the Intelligence Corps, where he studied Arabic and Hebrew. He was based in Cyprus from where he went to Beirut, working to find out about Israeli and Arab terrorist groups. His missions put him at considerable risk. He could have been taken hostage on several occasions. When he came home he was sent to GCHQ at Cheltenham. Then he went to Northern Ireland, where he became involved in the activities of the para-military organisations. His marriage broke up and he came out of the Army after fifteen years. Back in UK with his mind full of secrets he could never divulge, Anthony returned to a completely different career in engineering research in which he had once served an apprenticeship. All went well until a bright, over-zealous young manager sacked him. When I met Anthony he had not worked for a year and his GP, who had diagnosed depression, referred him to me. He had nightmares going back to his time in Beirut in 1975. The whole aspect of secrecy which had surrounded his work was a root cause of his problems.

However, Anthony was physically fit, he had remarried and wished to return to an academic life. He would not consider a War Pension for fear of revealing his secrets. There was also the stigma of mental illness which he wished to avoid. They had very little money so the Society sent them for a holiday in Bournemouth, paid for by the Not Forgotten Association. This gave him the confidence to enrol for a university degree course and he was able to qualify for a mature student's grant. When I last saw Anthony, he was half-way through his B.Sc. course in mathematics, and hoped to go into industrial research work. We never spoke about Beirut any more.

Anthony's case leads us on to consider those clients where non-military situations have led to mental illness; there have been many of them in recent years. I suppose traumatic situations have always left mental scars which only recently we have learned to recognise. Those affected by civilian situations would not concern me professionally unless the victims were Serving or ex-Service personnel, of which there were a few cases. I shall consider PTSD more closely in chapter 10.

Road accidents are a daily occurrence leading to trauma, and their effect on the Emergency Services has become a matter for concern. Military psychiatrists are at last being recognised as having the right experience, perhaps uniquely, to treat such cases effectively. We have become aware that disasters of all kinds large or small, natural or man-made, can leave behind victims suffering from severe trauma. These events are sudden, un-expected, frightening, and physical pain may or may not be a part of them.

Newspaper headlines in recent years have highlighted the capsize of the Zebrugge Ferry; disasters involving football spec-tators at Bradford, Sheffield and Heysel; there was a fire in the London Underground at Kings Cross; and on an aircraft at Manchester Airport. In earlier times there was the Great Fire of London and the Plague, as reported by Samuel Pepys. Even today there are great famines and earthquakes. Man-made disasters would include the sinking of the *Titanic* and the *Lusitania*, and the loss of the airship *Hindenberg*. If we agree that between a third and a half of all the people involved in a disaster may go onto develop PTSD, then if we add to this all those closely connected with the victims, we have a sizeable problem for our society.[5]

Smaller incidents can have equally sad consequences for their victims, and yet only reach the inside pages of the local news-paper, or not be reported at all. In chapter 7 I mentioned help being given by the RAF Benevolent Fund when they actually bought a house which they let to one of my clients. This was **Adrian** who was on his bicycle, inside the camp, on duty, when he was run down by an RAF truck in 1959. He was in hospital for a year with multiple injuries.

Adrian succeeded in recovering sufficiently to return to work. He became Warrant Officer, Chief Workshop Technician and he was eventually awarded the Meritorious Service Medal.

In 1980 he began to feel strange; he lost his temper, his legs failed him, his breathing became impaired, and his mental

capacity was reduced. Clearly he could not continue in a responsible position. Eventually Adrian was invalided out of the Service
and awarded a disability pension. We were able to help towards
getting him rehoused and stays at Tyrwhitt House gave his wife
a welcome break. Gradually his physical and mental faculties
improved as did his spirits. The Society was able on one occasion
to help him with his back pains by arranging a stay at the Royal
British Legion Churchill Rehabilitation Centre, where Adrian was
able to have much needed hydrotherapy. This was a successful
case, and an example which shows that not all my work was with
ex-Servicemen suffering from problems resulting from the stress
of battle.

Clients who came my way as a result of injuries which were not
from active Service included **James**, who was in the SAS, on a
parachute training course. His parachute let him down and this
resulted in injuries to his back and ankle. He tried to work until
his leg gave way in 1988. When I last saw him he was on crutches,
depressed and frustrated. His 40 per cent pension did not make
up for the loss of an active life. James said it would not be fair to
ask any girl to marry him, so he struggled to live on his own. He
did not wish to rely on his parents for help.

William was in the Guards when a marching column was
struck by a vehicle, in the blackout, during the war. He suffered
from spinal injuries but the trauma of this accident upset him for
the rest of his life. He saw no active Service, came out of the Army
and drifted from job to job until he finally managed to settle down.
When he retired, we were able to have his 20 per cent pension
increased. William had spent most of his life unaware of the possibility of even having his pension reconsidered. He was probably
afraid that they would take it away from him altogether if he
complained, especially when he was able to do some modest
work. I met many clients who had to be encouraged to stand up
for their rights.

Pete served right through the war as an anti-aircraft gunner in
the Orkneys. He was completely unscathed until 1945 when he
was accidentally blown up by a mortar bomb. His hands and arms
were badly deformed. It was assumed he would never work
again, and he was awarded a 100 per cent total disability pension.
He was trained as a painter and decorator at a Government

Training Centre, but he only managed to follow this occupation for about two years. Later he became a Civil Service messenger until he finally had to retire at the age of forty-nine. Understandably, when I met Pete he was mentally depressed by his physical problems. His wife was in poor health and had not been out of the house for many years. She said she would welcome a break from him.

Thus, trauma in ex-Services personnel is not just confined to those who have experienced ugly scenes in battle. The Emergency Services, although trained and experienced in facing the worst situations in road accidents, collapsing masonry and blazing buildings, have many cases of men and women who eventually discover that they have had more of this kind of work than the brain and body can withstand. In a category of their own have been those members of the Royal Ulster Constabulary who have coped with terrorism in Northern Ireland, alongside the armed forces, for the past quarter of a century. Not only has this entailed dealing with shootings and bombings, but also intelligence operations in pursuit of those involved and in anticipation of future outrages.

Fred Bishop was a Station Officer in the Brighton Fire Brigade when the IRA seriously damaged the Grand Hotel in 1984, in an attempt to kill members of Mrs Thatcher's Government. Fred rescued Norman Tebbit when part of the building was in a dangerous state, but this was only one of many traumatic encounters over a career lasting 29 years. He was awarded an MBE for bravery, but when he was invalided out of the Service he had high blood pressure, deafness, tinnitus, balance and sickness difficulties, back trouble, and understandably, depression. Fred was successfully rehabilitated towards the repair of domestic electrical appliances at Queen Elizabeth's Foundation for the Disabled at Leatherhead.[6]

Thames Valley Police and the Berkshire Fire Brigade continue to experience traumatic situations for their staff who handle serious road accidents on the M4 motorway; contact has been made with a view to obtaining help from the Psychiatric Department at the Royal Naval Hospital, Haslar. The Society has received expressions of interest in its work from some of the civilian emergency Services, but a lack of resources has always obliged them to confine themselves to treating ex-Service personnel exclusively. Where an ex-Service person has joined the police force, fire Service, or ambulance Service, and then suffered from trauma, help has of course been given, where possible.

Other categories of persons requiring help are the victims of crimes of violence, including rape, attempted murder, burglary and harrassment. Statistics in *The Trauma Trap* (pp. 75–7) reveal figures for notifiable offences in the western world running into millions every year. These victims, unlike the members of the Armed Forces, or the Emergency Services, never expected to encounter violence in their lives at all. Even so, when violence has occurred in battle, as we have seen, its severity, or repeated incidents, or the nature of the individual involved, can lead to a mental condition which, untreated, can ruin a person's whole life.

How treatment is administered, and its success and consequences, will be the topic of chapter 9.

Notes

[1] Schizophrenia, at that time, was regarded as hereditary and War Disability Pensions were not usually awarded in such cases. The official attitude rejecting this condition in WP applications was to be modified a few years later on.

[2] *Inside Madness* by Steven Hughes FRCS, *British Medical Journal*, Vol. 301, December 1990.

[3] L. S. O'Brien and S. J. Hughes 'Symptoms of PTSD in Falklands Veterans Five Years after the Conflict', *British Journal of Psychiatry*, 1991, vol. 159, p. 135.

[4] Hugh McManners, *The Scars of War*, p. 400.

[5] David Muss, *The Trauma Trap*, p. 33.

[6] Reported in *The Mail on Sunday*, 27 February 1994.

9

Treatment of the Mentally Disabled

Visiting disabled ex-Servicemen I encountered a wide range of mental illnesses. Here we shall consider some of the various forms of treatment, and detail the attitudes of some of my clients to treatment.

Perhaps the most serious comment was that there is no known cure for mental illness, and that psychiatrists are generally only able to give verbal assurance, with very little other support. Some clients were content to accept this, whilst hoping that an interview with a military psychiatrist would, at least, lead to a higher War Pension percentage, and more money to spend. Other clients regarded my visits as being purely social and an excuse for a cup of tea, and a talk about old times in the Forces. Then there were some who regarded Tyrwhitt House and the Society's other homes as just holiday centres, although very necessary, I must add.

I am not a psychiatrist, but in the course of my work I have found it useful to consider the wide range of my clients' conditions and place them into some kind of classification.[1] I shall also look at the various treatments that are available and consider their effectiveness. There are two basic types of treatment: social therapy and drug therapy. The one is designed to change a patient's attitude through discussion, the other to modify the internal mechanism of the brain (this latter treatment can also be attempted through electrotherapy). Both methods are not infallible and there are many questions which remain unanswered regarding the working of the mind. This is partly because surgical experiments on the human brain are not ethically desirable in the same way as the progress that has been made in the treatment of some physical conditions.

The most serious treatable mental conditions are schizo-phrenia and manic depression; the least serious are nervous anxiety and mild depression. PTSD lies somewhere in the middle of the spectrum of treatable conditions. On the one hand, there are the psychoses, where the patient has no insight into, or understanding of, his or her condition. Such patients include schizophrenics and those with affective psychoses with mood disturbance. Then there are the neuroses where the patient does have some understanding but he has some disorder of the personality. Further categories include the organic illnesses of the brain, and the dementias. Then there are subnormal states of the intellect, and cases of physical damage to the brain, some-times as a result of birth trauma. The neuroses embrace various states of anxiety, including obsessive compulsive disorders and hysteria. Personality disorders include psychopathy, addiction, deviency and social inadequacy. Some of these conditions are seen clearly to have been compounded by Service life. Sometimes it was argued, of course, that some of my clients should never have been allowed into the Forces in the first place. Some con-ditions are an aspect of social conditions or ageing, but nevertheless, as they were ex-Servicemen and women, it would have been morally wrong to cast them aside without attempting to offer some assistance, or comfort.

The nerve cells in the brain have interconnections, with elec-trical impulses going from one nerve cell to another by means of chemical transmitter agents. If they go wrong, then changes in the brain can occur. The use of drugs can stimulate or repress mental activity. A condition which is continuing is described as 'affec-tive'. A paranoia is a state of absolute delusion or an abnormal fixation or suspicion. A psychopathic personality is one having persistently irresponsible or aggressive behaviour. I have seen most of these states in my patients.

Any attempt to estimate the total size of the problem of mental illness can only be generalised. Fifty per cent of all occupied hospital beds are psychiatric; this is a legacy from the days of the old county psychiatric hospitals. Until thirty years ago schizo-phrenia was regarded as untreatable and the sufferers were put away together with some others who were, erroneously, thought to have this illness. There is also today a need for a large number of beds for elderly senile cases. Nowadays, psychiatric admissions are likely to be for only a few weeks and the number of permanent beds for psychiatric patients has decreased by 30 per cent since 1985 because of improvements in the treatment of schizophrenia. On the other hand, there has been some increase in hospital beds because today, on the whole, there are better chances of an effective treatment for many mental illnesses.

Care in the Community for the mentally ill was accepted in the UK following the lead of the United States, where it was felt to be wrong to lock away thousands of harmless people in large institutions when they could live out their lives in much smaller units, closer to the local neighbourhood. It is perfectly true that there were some who appeared, superficially, to be near normal in their behaviour. There were some politicians in this country who seized upon the idea of making huge savings for the NHS by closing the large county mental hospitals and selling off the land. There were, however, two basic false premises: the first was that these patients would happily assimilate within their new surroundings; the second was that follow-up facilities would be available locally.

Many patients were perfectly happy under the old system, in their own way. They were adequately fed, occupied and medicated where they had been for a long time, in rural surroundings where some work and physical activity were possible. It was not anticipated that public opinion would resent the presence of unstable characters walking the streets with nothing to do, little money and plenty of outside temptations for them to fall into. Furthermore, much of the expertise of the staff in the large hospitals was going to be lost or diluted so as to become non-effective. Some of the patients wandered off, leaving their homes in the community and ending up on the streets – thus bringing back the old problem of vagrancy the large hospitals were originally designed to prevent! The NHS might save on hospital beds but the new smaller units would be a burden upon local taxation, thereby shifting the problem of funding.

*   *   *

I discovered **Cedric** in a mental hospital where he had been since leaving the Army at the end of the war as an infantry soldier with severe battle fatigue. He earned a few pounds a week working in the hospital laundry. When Cedric told me something of his military record, we traced him and found that he was a War Pensioner with thousands of pounds of accumulated funds to his credit. He was visited only by his sister and she readily agreed that we should try him out at Tyrwhitt House. There, he was happy to join in the activities and, after two successful stays, visited by his sister, the idea was adopted that Kingswood Grange would be a suitable permanent home for him, where he would have some degree of independence together with an all-male ex-Service company. When a vacancy arose, Cedric was duly taken from the hospital. Fortunately for all, as it turned out, I had insisted that, should anything go wrong, the hospital would take him back. And I obtained this assurance from them in writing.

Quite inexplicably, Cedric began to turn very awkward, refusing to co-operate with staff and coming out with language none of us had ever heard him use before. His sister and I tried to reason with him but it was no use. We told him he would have to go back to the hospital and he did not welcome this news. Shortly afterwards, he physically attacked someone. He had to be urgently sedated, and I asked the hospital to take him back. Two strong male nurses arrived and Cedric was returned by ambulance under escort. He never wanted to see me again, the exercise was never repeated, and we can only assume that Cedric simply could not cope with his newly-found freedom. By his protests he was probably telling us that he was missing the security of the hospital. Had Cedric gone to one of those community units he could have been in even greater danger of injuring someone, possibly after falling into bad company.

In another part of the outskirts of London, the WRVS ran a clothing store for the needy which just happened to be not far from a large mental hospital which was being run down. Outside the premises, a group of former patients had been given a house in which to live permanently. Some of them would go round to the WRVS asking for clothing, and they were later discovered offering it for sale at a nearby public house. The ladies had to be asked not to serve these men and never to leave their handbags unattended in the shop whenever any of them called in. Such stories quickly spread throughout the local community and the ex-patients certainly did not fit in.

Psychiatric hospitals have offloaded many of their patients into community houses, or into psychiatric and geriatric units set up within the general hospital. Problems arise here because of the lack of space for the occupational and physical exercise facilities formerly enjoyed. The old mental hospitals have retained a few schizophrenic patients from the days before there was an effective treatment for this condition; there might also be a few dementia cases which need continuous care and have nowhere suitable to go. One serious aspect of moving psychotic patients cannot be emphasised enough, and that is the problem of disorientation should some of them be moved to fresh surroundings. Cedric's difficulties were an aspect of this. In the case of my own father, my mother, herself an invalid, successfully looked after him through many years of senile dementia. When he had to be moved to hospital for a physical illness, he died there within a matter of a few weeks, in total confusion. The poor chap thought I was his brother who had died many years earlier.

The psychiatrist's out-patient clinic today is more or less equally divided between four basic types of patient. There are the psychotics, schizophrenics, and manic depressives. Secondly, the psychoneurosis cases with anxiety states and depressive reactions. Thirdly, the psychosexual problematics with impotence, frigidity, deviation or homosexuality. Lastly there are the personality disorders, the inadequate, social dropouts, unsuccessful suicides and disturbed epileptics. Most are on medication of some kind.

Whilst it is true that we cannot cure all of the mentally ill, many conditions can be greatly improved nowadays. This is more than can be said of some physical conditions which exist today, such as cancer or AIDS. These are facts which cannot be disputed and which I repeat to all those who complain that psychiatrists are not effective. There were 145,000 psychiatric beds in 1950, 110,000 in 1970. Psychiatric hospital admissions were 60,000 in 1950; there were 175,000 of them in 1970 but most of them were for a few weeks only. Forty million working days were reported lost through mental illness in Britain in 1972 with £45 million pounds paid out in sickness benefits for the same causes.

The most significant reason for hospital admission today is for some kind of temporary care: 95 per cent are informal or voluntary patients. The other 5 per cent are for those without insight or understanding, and are usually admitted at the request of the family or friends. Often these individuals are a danger to themselves or to other people. It is interesting to observe that psychiatric cases were first recognised as being a national problem by their inclusion, as an afterthought, in the 1948 National Health Service Act. The county mental hospitals had previously been the responsibility of the local authorities.

The Mental Health Act of 1959 was a milestone, later modified by the Act of 1983 which made mention of abnormal aggression in patients who can be outside the Criminal Law if no indictable offence has been committed such as murder or grievous bodily harm. From 1959, patients could be detained if there was a danger, on application by a social worker or next of kin, with the support of two doctors. Under Section 2 of the 1983 Act a period of twenty-eight days is specified for detention in hospital; under Section 3 such a period can be extended to six months. Those patients who are detained under the Act are known to the professionals as being 'sectioned'. A case of attempted suicide can be admitted for 72 hours initially, and a nurse can hold a patient for up to six hours pending the arrival of a doctor. A police constable can take a person in need of care or control to a place of safety, to be seen by a doctor, on his or her own authority. Appeals against detention are allowed for under the Rules and there are slight differences in Scotland, Northern Ireland, the Channel Islands and the Isle of Man, where there are separate legal systems. I have attended such appeals and usually it is the opinion of the psychiatrist which prevails, sometimes to allow the appellant to go home. On occasions I have been asked to arrange for such a client to be taken into Tyrwhitt House, as a condition of his discharge from hospital.

The Court of Protection exists to manage the affairs of someone unable to look after himself. (Incidentally, a Member of Parliament would lose his or her seat after an absence of six months in a mental hospital; no such rule applies to the House of Lords!) It was the 1959 Act which created the situation of having voluntary patients; the 1983 Act reinforced the rights of these patients.

Manic depressive psychoses have been treated by the use of opium, bromide and the barbiturates. Convulsive therapy, the use of insulin, electrical stimulation and brain surgery to induce personality change, were also previously highly regarded. However, modern methods of drug and behaviour therapy have led to a decline in the older methods over the past forty years.

The 1950s saw great advances in drug therapy. It was found that the hospitals could unlock their doors and only a few ran away permanently. There was an increase in part-time day care. The drug chlorpromazine became available in the early 1950s for controlling psychotic behaviour, but there were some side-effects which produced a kind of Parkinsonism in some patients. Now there are safer drugs requiring smaller doses, fewer side-effects and they are longer acting. Similarly, anti-depressant drugs were developed in the form of mono-amine oxidase inhibitors (MAOI), as well as the tricyclic anti-depressants for the affective psychoses – again with fewer side-effects. The anxiety control tranquillisers now used are the benzodiazepines. All this progress in drug therapy has made it possible to integrate mental care within the general hospital with some day hospital care.

Most NHS psychiatrists are overworked and many patients who need psycho-therapeutic techniques are neglected in favour of using prescription drugs. Some of this therapy has been simplified, but if we consider Freudian psychoanalysis, with conditioning and the interpretation of dreams, there could be a pattern of up to two years' attendance, which the NHS could never afford.

Psychotherapy includes counselling, behaviour therapy and psychoanalysis. The relationship with the therapist is very important, but there are dangers in the patient becoming dependant upon the personality of the therapist. Younger clients with mild neuroses have been found to benefit most. Behaviour therapy can be used for phobias, or in making a pleasure unpleasant – as in trying to cure alcoholism – but the patient must co-operate if he or she is not going to relapse. Psychotherapy influences the patient's thought, helps him or her combat repression (the brain's means of concealing a mental defect), and promotes a coping strategy to deal with neuroses. It cannot help cases of dementia, the psychoses or schizophrenia. There is no legal requirement for the therapists to be trained, and the patient's co-operation is always needed.

Supportive psychotherapy is counselling or guidance. This is used by some clergy and social workers, and it can lead to an exploration of the underlying cause of the condition. Successful therapy can lead to a change in life-style and the development of insight. The latter is most important. Most patients under the NHS are offered medication with, perhaps, some short-term counselling from a social worker, nurse, or a clinical psychologist; there might be a case conference for back-up in case of continuing problems.

Group Therapy was first practised in the First World War. Further advances were made in this form of therapy after the Second World War, at Northfield and Mill Hill, as narrated in chapter 5. The group can provide its own dynamic force with the leader or therapist interpreting the group's reactions, sometimes acting as a catalyst, or as a passive but subtle onlooker. Eight patients might assemble, say, all with similar problems such as alcoholism or a phobia. There would be weekly meetings over some six months, and confidentiality is vital. Some therapists might organise psychodrama where the group acts out the source of some trauma. A good question to start the discussion might be 'How will we know when we have got what we are coming to the group for?' It is good to target all the problems first of all. The group might begin with everyone on tranquillisers at first before they are gradually weaned off them, perhaps together.

There are always the dangers of some members of a group adopting what they think are the required attitudes or giving what they think are the correct answers. Very many of the questions will have no universally correct answer, but I have found some ex-Servicemen very prone to behave in what they would regard as a correct manner when being interviewed. This is perhaps a legacy of their military training when practically everything had just one correct answer and the basic philosophy was always to do as you are told first and ask questions afterwards (if you must!). Many a regimental medical officer has encountered this problem when a soldier was trying to return to duty before he was fit enough. I met it in men who thought that I had arrived to have them taken away to the old asylum.

Those who have not benefitted from group therapy may need a one-to-one course with a doctor. Such a short course has certainly worked when I have sent a problematical client to see a psychiatrist at a military hospital for a series of interviews, especially in cases where a local NHS consultant, or the patient's GP, have been unable to help at all. Unfortunately military hospitals are closing down all over the country, so this solution to the mental health problems of ex-Service personnel may be unavailable in the near future. The military psychiatrists, a scarce breed, may well become scattered. Their expertise might well be lost to us for cases where it would have made all the difference.

Within the group, adverse reaction to any psychotic behaviour, such as hearing voices, should be carefully channelled. The group might be able to help by discussing where such voices might come from. Many of them will never have considered this before in their lives. Thus, the group should aim to give the patients some insight into their condition which will lead to its control. Hopes can be shared as individuals are led away from the selfishness which typifies many neurotic and psychotic conditions. In this way the group can beneficially motivate the individual. The results of group therapy can improve the individual's attitude to the group itself. This can be very important should we be attempting to encourage the patients to return to work for a living. Those who are totally selfish, and sorry for themselves, will probably have lost all their friends and are unemployable because of this. If patients share responsibility for a group task, this can lead to an increase in independence and the end of being sick.

Electroplexy is a generic term for the use of an electrical current to induce a seizure discharge within the brain itself. The most common treatment is electro-convulsive therapy (ECT) and this produces a seizure discharge within the central nervous system so as to trigger off activity. It has been used in cases of threatened suicide, but the depressive psychoses, on the whole, have been the conditions treated most successfully by ECT. The treatment is highly controllable through the strength of the electrical impulse used, its length of operation, the frequency of visits and the length of the actual course of treatment. Quick-acting anaesthetics are used together with muscle relaxants before the treatment commences. Only a few schizophrenic cases have been found to benefit, but the manic psychoses can be treated more rapidly by a combination of ECT and medication. It can also help modify attacks of epilepsy. However, the results of ECT can never been guaranteed to be permanent, but, when it does work, it can be more effective than drugs. Cases of schizophrenia where ECT has worked have been those where the phenothiazine drugs have been used over the same period.

Doctors and professionals are not certain what ECT actually does, and there can be some side-effects such as headaches or temporary memory loss. However, it has been shown to be safer than aspirin or some of the anti-depressant drugs. It must be said that no deleterious effect on the functioning of the brain has ever been established as a result of ECT.

*   *   *

Geoff was an ex-Royal Naval Chief Petty Officer, married with two teenage daughters, and a patient in one of the old mental hospitals not far from where they all lived. He was permanently in a wheelchair having had his back broken when the muscle relaxant accompanying ECT, administered at a Service hospital after the war, did not work. He was awarded 100 per cent War Pension and discharged.

He had tried to live at home, but all he wanted to do was to lie in bed all day. He was frustrated, could be unpleasant, and the girls were afraid of him. Geoff was taken back into hospital. When I saw him, he was determined to be with his family again, but they were reluctant to have him home. When the hospital decided to discharge Geoff, the War Pensions officer and I made the strongest possible representations, but to no avail. He was put into a flat in the same town as his family, with a promise of sufficient support from Social Services.

This support never materialised and Geoff was left to fend for himself, hobbling about the flat on crutches. I never saw him again. When I called, no one came to the door; a neighbour told me he had no concept of the time of day, so he slept in the daytime, got up in the night and wandered about outside when the shops were all closed. When it was found that he was not feeding himself properly, he was taken into the psychiatric wing of the local general hospital. Geoff was clearly a casualty of an ECT which had gone wrong, together with a blatant display of the inadequacy of Care in the Community. When I complained, all I received from Social Services were promises of increased support from them. I believe his wife was later able to go out to work and they were divorced. From what I knew of Geoff's behaviour, I could understand why she mainly wished to get on with her own life and look after the girls.

One of the problems highlighted by Geoff's case was that caused by the reorganisation of Social Workers in the 1970s. Mental Welfare Officers, who worked exclusively at the old psychiatric hospitals, were amalgamated into a general social Service system which embraced those working with children as well as with the elderly. In theory, all the social workers could do what the mental welfare officers used to do, and this created problems arising from their inexperience. One idea was that the social worker should form part of a multi-disciplinary team, with the psychiatrist, community psychiatric nurse (CPN) and the GP all working together for a complete family which has a mentally sick member. This was a laudable concept but none of the participants really had sufficient time to devote to it. It has been estimated that 25 per cent of the work of GPs has some psychiatric bearing; yet there have been problems (to be discussed later in the book) when physical illnesses have been dismissed as psychosomatic and psychiatric problems have not been diagnosed at all.

Brain operations (leucotomy) are rarely used nowadays. Previously they were used in cases of severe mood disorders, chronic anxiety, suicide, obsessive compulsive disorder, Parkinson's disease and epilepsy. They were used in these conditions where there was no suitable alternative treatment. However, adverse public opinion prevented some possibly worthwhile operations from taking place. Experiments on the brains of animals are practically impossible to evaluate and, understandably, public opinion is against such things.

The lack of ex-Service support, the reorganisation of the social workers, and the Care in the Community Act of 1990, implemented in 1993 were all clearly shown up in the case of **Steven**. He was a young man who suffered head injuries in a road accident in Germany and whose mother came to me for help. The difficulties facing a soldier discharged after falling foul of Military Regulations, not entirely his fault, were also brought out by this case, which could have been better resolved. Steven and his comrades, under the command of an NCO, took a military vehicle to go into the nearby town, unofficially. They were involved in an accident for which the NCO was court-martialled, and reduced to the ranks. The others, not severely injured, were reprimanded. Steven received head injuries which led to the onset of PTSD, from which he had nightmares after his discharge. The Army gave him an adverse record of conduct although he had given many years of satisfactory Service before the accident.

When I saw him, Steven was unable to work or manage his affairs on his own. Shortly before the DSS changed the rules, the Society was able to give him a very satisfactory stay at Tyrwhitt House, even though the case for a War Pension was going to be difficult under the Rules, perhaps impossible.

After this period of nursing, his mental condition improved sufficiently for Steven to start a course at the local technical college with a view to qualifying as a civilian in the trade he had whilst in the Army. By then he was in a flat on his own, with a social worker helping him manage his finances. He had no idea of the value of money – a common failing among mental patients. It was clear that further progress might be made if the stay at Tyrwhitt House could be repeated. There, the Society's staff might report on him further as part of an appeal for a reconsideration of a War Pension case, previously rejected, under the Rules.

At a case conference, which was now necessary if the Local Authority was to sponsor such a visit, and now required under the new Rules from April 1993, his case worker implied that there was nothing mentally wrong with Steven that they could not handle; she calmly accepted that he might never work again, although he was only in his twenties. Clearly, they had no idea of the possibilities open to Steven as a disabled ex-Serviceman. Someone not at the meeting had already decided that they were not going to pay for him to receive treatment. We were left with no help at all beyond keeping him amused in his flat with a social worker looking in occasionally. Steven gave up the idea of continuing his course at college. His GP was not very helpful either. When I saw Steven for the last time, he had been persuaded by someone that the Society was really only there to help elderly war veterans and as he no longer wished to co-operate with me there was little else I could do. His mother was disappointed and said she felt betrayed all round whilst admitting that I had tried my best for her son.

When it comes to complaints by ex-Servicemen, the Ministry of Defence is very difficult to sue and it is only recently that such claims have been possible at all. The Royal British Legion continues to campaign for a Ministry of Ex-Service (or Veterans') Affairs with a senior Member of the Government able to represent our clients in Parliament. Such a ministry is in existence in Australia, New Zealand and Canada. One wonders whether such a ministry would have been able to obtain a redress of grievance for Steven by way of an amended Terminal Report, to help him obtain employment, and the possible reconsideration of his case for a War Pension.

Some of my clients have shown aggression and it was some-times wise to place myself between the person I was visiting and the door. It might be a problem if my visit took place towards the end of my patient's monthly injection cycle, when the calming effects of his medication were wearing off. There were some clients who would suggest the best time for me to see them. Indeed, the regular visits by the CPN were a lifeline to several of my clients who would have been locked away in former times.

Hospitals, as a rule, do not lock patients away today, but prison sentences with medical care are available at Broadmoor and other secure hospitals. Secure accommodation for prisoners on remand, with medical care, and for those otherwise dangerous, but who have not committed a crime, is not now sufficiently available. The question of criminal responsibility is a legal one, not a medical one; the psychiatrist advises as to the patient's condition but it is the Court which decides the sentence. When patients are discharged from special secure hospitals, having completed their sentence and responded to treatment, the decision to release them is made by the Home Office. There can be the danger of a relapse and the population outside can be hostile to such a release. Follow-up care is absolutely vital, yet not always adequate. Every time a psychotic personality commits a crime the news hits the headlines, a life may have been lost and the public demands more secure accommodation for these patients. We have not solved the problem of dangerous patients who have been discharged from psychiatric hospitals or prisons. A person on a suspended sentence, with the involvement of a psychiatrist in his treatment, can still be a danger to all, especially if no one sees to it that he takes his medication without fail. A hospital, other than one of the secure hospitals, will not wish to take in a patient as part of his sentence when all the others are voluntary patients. This is a serious problem and an outcome of the decline of the old system of large secure mental hospitals; the problem was probably not foreseen.

There are big differences in the needs of the long-term patients in mental hospitals and the shorter-term ones in the general hospitals with a psychiatric wing. Specialist units with plenty of space for activities are still needed, although some of the old county mental hospitals have been, and will be, closed. Day hospitals at district general hospitals must expand.

The Community Psychiatric Nurse (CPN) is the lynch-pin of care in the community for the mentally ill. Some nurses have come from the former staff of the large hospitals, now run down, but allowances have to be made for travelling between the patients in their own homes or day centres. Immediate action is vital should anyone miss their regular medication or wander away from their last known address. I realise there are those who fear the imposition of a 'prison state' regime, but we are not dealing here with the average citizen. Some of my time was lost through trying to find the whereabouts of patients no longer at their last address.

The Society's Scottish Office informed me of the likely arrival of one of their clients in Coventry, which was then in my area. I had an address, which was a hostel for needy men in a terrace of old houses right opposite the Rolls Royce factory. The premises were not palatial but adequate, and the food was basic but good. After a few attempts, I was able to interview **Jock**. He was polite and friendly and told me how he had walked from Glasgow and would eventually make for London, on foot. He valued his freedom to wander and lived on a small War Pension; he would not disclose how he collected it. Perhaps he had a relative or a forwarding address somewhere. I gave him a pair of new shoes for which he was grateful, and that was all I could do. He refused any mention of a small gift of cash. He was made welcome at the hostel but was suspicious of making any permanent arrangements. He described himself as an independent Scotsman and any mention of Tyrwhitt House or Kingswood Grange was out of the question. I never saw him again.

The trend towards reducing the numbers of long-stay beds will have to be reversed if only because of the increasing numbers of elderly patients who need care and are living longer, sustained by modern medication. After closing a hospital, a visiting crisis team of experts could put some patients off who really only wish to discuss their problems with one person to whom they can relate. There is also the danger of low morale among staff in hospitals due for closure. The loss of the benefits of living in the community of the large hospital has probably not been fully appreciated; then there are the problems of moving patients who are set in their ways, probably elderly, confused, perhaps of low intelligence. There have been some early deaths following such moves in recent years. To the patient, the small unit can feel more isolated than the large ones were; and some patients have wandered off without trace. Increases in the numbers of medical and nursing staff, proudly announced by the politicians, have produced increases in the numbers of inexperienced staff working shorter hours and this, in turn, has added to the confusion of some of the patients.

Social workers have become more generalist, replacing the old mental welfare officers. And some psychiatrically trained social workers have gone into administration, promoted as the Service has expanded. The loyalty of the social worker is now to the Local Authority, not to the hospital, as was seen in the sad case of my client Steven.

A new profession, that of clinical psychologist, has arrived and he or she might specialise in the study of neuroses. My client Roy, mentioned earlier, benefitted greatly from interviews with a lady clinical psychologist. She gave him renewed confidence at a time when further admissions to Tyrwhitt House, under the recent change in the Rules, seemed unlikely. At the very least, she was able to give him a very welcome alternative source of support.

In recent years there has been an expansion of occupational therapy. However, the occupational needs of the mentally ill patient can be quite unlike those of the physically handicapped. The two types of client can seldom be treated together. In some cases, half the battle has been to persuade a patient to do anything at all, if he is severely depressed or has an impairment of the brain which affects his co-ordination. The War Pensioners' Welfare Service has always had handicrafts officers who paid home visits to provide materials and knowledge to those who wished to occupy themselves. Here again, much was done for the physically disabled pensioners, and there is an annual nation-wide exhibition of their work held in London. The occupational therapy departments at Tyrwhitt House and Hollybush House, in Scotland, have always been highly successful. The secret has probably been to take each visitor along just to see the others at work. As a result, and with no pressure at all, and plenty of cups of coffee or tea, many patients have become involved in some of the wide range of crafts on offer. These places have progressed a long way from simply basket-weaving and drawing pictures.

Research continues, and the better understanding of the mechanisms of the mono-amine transmitters in the brain could improve treatment for schizophrenia. There is also on-going research into the anticipation of the onset of this illness. Moreover, better follow-up of discharged patients can lead to a six weeks' stay in hospital being the norm for schizophrenia. However, it will be necessary to make some allowances for some who might require further admissions should they relapse.

Regarding the affective psychoses, evidence of a genetically determined bio-chemical disorder in the brain, causing them, could lead to the arrival of a more specific antidote than is available at present. The need for electroplexy could be further reduced.

The number of cases of psychoneurosis continues to increase, as does the number of would-be counsellors. Tinkering with other members of the family, unless very well done, can cause further distress to them. I found families who appreciated a chat and I learned from them much about the patient; but I seldom offered them advice, if only because I felt I could never know all the facts about their circumstances. If they came to me with questions, then I would try to advise, even if it was only to contact a specialist in their type of problem. I had plenty of this kind of information available via Society headquarters.

The benzodiazepine drugs are on the decline, but no effective drug seems ever likely to be free from some side-effects or the

Work and relaxation at Hollybush House, Ayrshire, opened in 1985.

danger of addiction. The effect of a drug on a patient may depend on a number of factors, including size of body – a large patient may require a larger dosage for the drug to be effective. One of the problems with the elderly is that they become more frail and doses of medication need to be progressively reduced. This factor is sometimes overlooked. Some drugs affect other parts of the body on the way to the brain, such as the balance of acid in the stomach, or they may be absorbed by the liver. Alcohol can do this. A drug with a long period of dispersal can accumulate, leading to overdosage. The result might be a worsening psychiatric effect, bringing about a situation where further drug administration might be incorrect. Sedatives, tranquillisers, anaesthetics and alcohol can all interact. The doctor may not be aware of all that the patient is taking, especially if he does not co-operate or understand. And indeed there are patients who are like this.

Better use of group therapy might allow more clients to be reached, but factors of time and expense militate against the NHS. Thus, the voluntary agencies, with appropriate staff, trained and experienced, might be of help here.

Senile dementia tends to be considered an incurable problem. Yet, the establishment of its metabolic cause might be possible in the future, and this would constitute a major breakthrough. Today, if we can recognise it early, some medication can delay its onset and possibly modify its longer-term consequences.

The young disordered personality requires security; some secure units are being provided because the special hospitals are full. Each region should have such a unit, as they have for alcoholism, drug addiction and epilepsy. There is every likelihood of better future use of psychiatrists in the discovery and treatment of psychosomatic illnesses which are not easily diagnosed today. Both these aspects will be considered later.

Note
[1] See the chart on p. 215.

10

A Wide Range of Cases

When I began to work with the mentally ill, I had little idea of what my patients would be like, except that most would be unusual in their behaviour and reactions, some possibly hostile or even violent. As time went by, I found I could categorise them along fairly clear lines, particularly between those who wanted or appreciated help and those who did not. It was obviously easy to turn one's back on the latter and sometimes, as when a client persistently refused to see me, little else could be done. Whenever our paths crossed, busy GPs would tend to agree with me. Some clients declined to see me after it had been explained that it was not going to be possible to improve their condition. Some were reluctant to reopen old wounds, some clung steadfastly to their feelings of independence, some were hostile to those who had been sent by Social Services or other official bodies if they had clashed with some Regulation or with the Law, or with some neighbours who may or may not have been well-meaning. Those who turned their backs on me were, I felt, often those most in need of some kind of help – but not always financial help.

The chart of psychiatric illnesses and conditions shows, in brief, the extent of the field of medicine covered by psychiatry. Is it any wonder that psychiatric problems take up a sizeable proportion of the GPs' time and the NHS budget? Broadly, we are looking at two sets of patients, those who have insight and those without insight (those who do and those who do not understand, realise, or accept that they are ill). A psychosis is a serious condition of the mind where insight is frequently not present, and this condition will dominate the patient's whole mode of thought, his life and actions. I met clients within all the main categories of illness detailed in the chart, but I propose mainly to mention those who formed a major part of my case load.

Psychiatry has been defined as that branch of medicine which deals with conditions displaying a disturbance of thought, emotion, and consequently, behaviour. The psychoses include the schizophrenias, paranoid states and paraphrenia. I should add that many war veterans were diagnosed as schizophrenic simply because their behaviour was odd or not understood, when, in fact, they were neurotic – which is in fact a different type of illness. Schizophrenia used to be erroneously thought of as a condition where someone had a split personality; the condition tends to be permanent but the patient can have episodes of rationality. (A split personality today would probably turn out to be evidence of a bipolar manic depressive condition.) The root cause of schizophrenia is not known but it could be genetically based. There are biochemical disturbances in the brain which cause distortion and fragmentation of thought; these are possibly triggered off by environmental factors. This latter idea has been very important for my patients who, diagnosed as schizophrenic by the military authorities, have applied for War Pensions. Such applications have tended to be more successful in recent years.

When I first became a Welfare Officer, I found that schizophrenia was generally regarded as hereditary and thus could not be connected with military Service (as in the case of Darren detailed on p. 163–4) In recent years, the idea has become accepted that this illness can be triggered off by stressful events. I have met a number of individuals where schizophrenia has been regarded as an accepted condition, aggravated by war Service, rather than directly attributable to it.

In the schizophrenic patient, thought is disordered; emotional behaviour can be abnormal, with a lack of energy or drive, and with some withdrawal from the real world. Schizophrenics can have delusional ideas and hallucinations, sometimes in the form of voices which tell them to do things. The delusional ideas can take over a client's life-style; it is then said to have become obsessional or paranoid. A typical case can develop into one with abnormal suspicions or a persecution complex. Some of my clients firmly believed they were being spied upon by the Russians during the Cold War, or by the Police or a Government

Psychiatric Illness

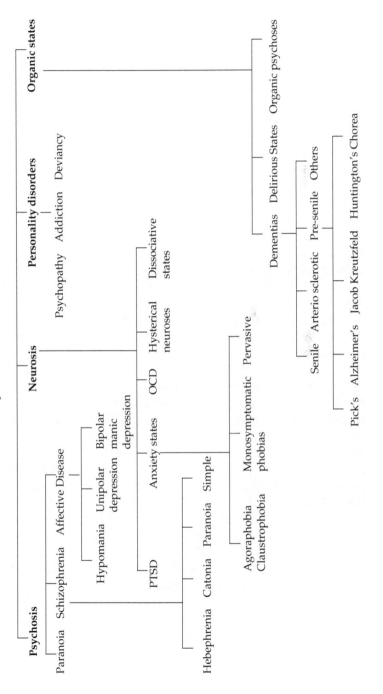

department. A schizophrenic can have sudden outbursts of activity or changes of mood. Then there might be imaginary friends. I do not propose to discuss all the various ways in which this illness has been classified, but one thing which all my clients had in common was that they lived in a world of their own making and which, to them, was totally real. Because of this, one could say that they were selfish, perhaps unwittingly. On the other hand, some were most friendly towards me; they regarded me as a friend just so long as I did not try too hard to pry into their fantasies. They would just never accept that any of their ideas was wrong.

*   *   *

My first case was an aggressively selfish man, and many hours were spent on trying to help sort out his life-style, but to no avail. The main problem was that, in speech and manner, he could appear to be normal; so, many were taken in as they listened to his tragic stories of misfortune. **Edmund** was ex-RAF and the Benevolent Fund had taken pity on him, helping him obtain a partial mortgage on a new house, through a housing association. He had a 20 per cent War Pension for schizophrenia aggravated by the Service, which we thought he was lucky to have. However, Edmund's life-style never fitted his means and he believed that the world owed him a grandiose mode of living. Whenever he felt gloomy, he would go out for an extravagant meal, or buy some expensive new clothes, or indulge himself in records or CDs of classical music. He paid via credit cards which he could not afford, and he ran up debts with several companies. Edmund had changed his address every time he imagined his neighbours were threatening him. Then there were the bills which kept on following him around, until he hit on the idea of changing his name periodically. In this way, when Edmund arrived at a new address, the Electricity Board thought he was somebody else and connected him without Edmund having to first pay the arrears he owed. His debts were eventually written off by the Board.

When I first saw Edmund, he insisted on talking with me in the car because he was certain the neighbours would be listening through the wall. He stated that he wished to preserve his independence, but I told him that he could not possibly expect to continue spending money he did not have. At this point, a doctor arrived to assess Edmund for the Mobility Allowance. Edmund had read somewhere that it was another allowance he was not receiving. The doctor promptly told Edmund that he was as mobile as he (the doctor) was. This made Edmund angry. 'Do you know your trouble? You're like him (meaning me). You're prejudiced against me!'

I had arranged to be present when the Community Psychiatric Nurse (CPN) called to administer his monthly injection. The nurse offered to control Edmund's finances until it became quite clear that things were never going to work out. I asked him where precisely would he like to live. 'I'd like a cottage in the country'. 'Wouldn't we all!' replied the nurse. Later, I had a phone call from Edmund, at Buckfast Abbey. 'What are you doing there?' I asked. 'I'm seeking sanctuary from my persecutors.' 'And what does Father Abbot think about that?' 'He says I have to leave by the beginning of next week!'

So, yet another of Edmund's fantasies did not work out. Much later, we had him in one of the Society's homes; this was until he had to be asked to leave because he was 'borrowing' money from some of the others and not paying it back. I last saw him at his sister's. She had taken him in after she lost her husband. Edmund was merely taking advantage of the situation, paying her very little in housekeeping money, and spending all he had left on luxuries. The dear lady had very little money, but my remarks about paying her more just fell upon deaf ears.

It is hard to see what the Society, the RAF, War Pensions, or Social Services could have done for Edmund when his fantasies were out of control. I suspect that he did not take his medication, except when supervised. I am sure Edmund thought he was normal for most of the time. Years ago, in a large mental hospital, he might have lived a reasonable sort of life at no danger to himself or to others. As it was, he simply took advantage of the law as it stands, and which was inadequate to have him permanently supervised. Edmund always believed that he was the only one in step. I know the local hospital took him in occasionally but, as soon as he felt better, he would discharge himself, which he was entitled to do.

Cyril was a completely different type of schizophrenic. He was called up in 1942, straight from grammar school. He served with the Eighth Army, before being brought home for the D-Day landings. He was a forward gunnery observer. Cyril was commissioned after the war. On leaving the Army, he entered university, became ill after two years, and was diagnosed as schizophrenic. He was in a psychiatric hospital for some time, and never completed his degree.

When I met Cyril, he was living alone and he had become a student with the Open University. He had tried teaching in a preparatory school but without success. I found him always charming and friendly although he was the victim of his own fantasies which ruled his life. He was convinced that everyone he met worked for MI5, including his doctor. They were all filing reports on him but he never divulged where these reports went! When I took Cyril to a War Pensions Tribunal, in an attempt to obtain some recognition for his illness, he did not answer any of the questions put to him; instead he 'blew it' by talking ad nauseam about the Russian agents who were watching him. The Tribunal was unable to make any connection between his illness and his military Service, and he prevented me from asking our pre-arranged questions by taking over his own case. I am sure he believed he was a kind of 'Perry Mason', the Chief Attorney for his own defence. On the other hand, at the time of the Tribunal, schizophrenia was not generally regarded as possibly connected with war Service.

Once when I called, and Cyril was always pleased to see me, with tea and fresh cream cakes, he repeated his claims that he was being continuously spied upon. This time it was by British Telecom. I looked under the bed, behind cupboards and wardrobes, and I pronounced a total absence of any bugging devices in the house. At this Cyril said, 'I could have told you that. It's all done by satellites these days and you can't see 'em!' When he travelled by train to see his university tutor, he believed that British Rail had hidden cameras in all the carriages. Cyril also believed that some of his university work had been stolen by a spy. Nevertheless, he passed his examinations and was awarded an M.A. degree. When I suggested that his papers had been merely lost and found again he quickly dropped the subject.

When my area was reorganised and I no longer had Cyril as a client, I received a telephone call from a very worried hospital administrator. Cyril had been caught trying to attack the television cameras in their car park. He had been convinced that he had at last discovered who it was who was spying on him. He was led away by the police, and was later interviewed by a friendly psychiatrist, after I had explained about his case.

My third schizophrenic patient was another soldier who had served right through the Second World War. He was for ever moving around, had no friends and was badly in need of clothing. When I first met **Daniel**, he was a seaside car park attendant. I promised him some warm items for the winter but, when I called again, he had disappeared. His landlord told me that Daniel would have been welcome to stay as he was well behaved and a veteran of the Burma campaign. What neither of us knew was that Daniel responded to voices he heard inside his head.

I had told Daniel about the Society's homes but it came as a surprise to learn that he had hitch-hiked to Milner House from South Wales, to our hostel in Leatherhead. The Warden was able to accommodate Daniel and he appeared to settle down. Daniel seemed to be less drawn and anxious and they found him odd jobs to do.

I next saw Daniel in a Social Services resettlement centre where they were trying to find work for him. The Centre Manager later reported to me that he was in hospital. When I arrived there, he had moved to the Royal British Legion home at Rhayader in Mid-Wales.

Crosfield House was, and I am sure still is, a very comfortable home. I often described it as a five-star hotel. Daniel was very happy there and wanted for nothing. When I mentioned clothing, he showed me a full wardrobe (fitted, of course!). He had complained of pains in his stomach and they put him on a special diet. He only had to ring for 'room Service' and they brought him anything he wanted. However, his last words to me were, 'Come the Spring and I'll be away again'. Daniel never saw that spring. His stomach pains turned out to be cancerous and his life ended rapidly. His funeral was attended by comrades from his old regiment. Daniel would have appreciated this.

Other types of psychoses, where the patient has no insight

into his illness, are those described as 'affective'. These feature a continuing emotional state, usually of depression or elation. If the condition is bipolar, it results in depression and elation, sometimes changing moods in a matter of seconds. As with schizophrenia, there is clear evidence of a genetic predisposition and the condition may be triggered off or worsened by the environment or by an event. So it was that these clients were able to qualify for War Pensions more often than the schizophrenics.

Depressive clients encounter morbid thoughts and feelings of unworthiness; it could be easy to dismiss these conditions as based on an inferiority complex and to tell them to 'snap out of it' – but that attitude was a serious mistake, as I was to discover. A large proportion of my cases were manic depressives and how much insight some of them had is not known. They all knew they had uncontrollable gloomy thoughts, but many just accepted them and took their medicine. When they did not, things became worse for them.

When researching for this project, one of the most interesting books I discovered was, *Depression, and how to survive it* by Spike Milligan and Professor Anthony Clare (Arrow Books 1994). Spike has been a friend of the Ex-Services Mental Welfare Society for many years and I remember him doing a fund-raising broadcast on BBC Radio 4. This raised not only thousands of pounds, but also produced hundreds of cries for help from ex-Servicemen all over the country, all of which were answered.

The book revealed that every year a considerable number of people commit suicide as a result of depression, and I will be telling the story of one client who did just this. Many individuals receive no therapy at all – they fear shame, discrimination and isolation. The immediate causes of this may be marital, bereavement, or unemployment, or conditions encountered in the Services. Few accept depression as a physical illness, while antidepressant drugs may have been shunned because they were thought to be addictive.

Spike Milligan suffers from recurrent bouts of manic depression. He has had hospital treatment, including ECT, but he still encounters a bad patch from time to time. He can be depressed for months at a time, when he tends to withdraw from everything. Both his parents had some history of a neurotic temperament.

Spike describes his depression as life without colour, food is taste-less and human feelings disappear into a kind of paralysis. Although depression set in around 1953/4, he attributes it to events in 1944, when he was blown up by a mortar bomb at Monte Cassino. This probably triggered off something latent in his personality. He was to lose faith in human nature, and sometimes he felt depressed even when things were going well for him. It is popularly believed that mental illness correlates with violence and indeed Milligan once threw a fit at Broadcasting House. Some patients will react violently if crossed and I was always careful never to confront my patients too strongly. Some of them felt guilty for being a burden on society. On the other hand, the condition known as mania can lead to talkativeness, hyper-activity, buying sprees, foolish investments and strange flights of ideas. Milligan's illness can go from one extreme to the other and is therefore classified as 'bipolar'.

The Mental Health Foundation estimates that six million persons in the UK suffer from some form of mental illness (10 per cent of the population). Most of them will consult their GP for the physical signs of depression. Doctors, for their part, are reluctant to diagnose mental illness because of the stigma attached to it. As a result the real problems are seldom resolved. Women are twice as likely as men to suffer from depression.

Depression is treatable. It may begin in the late teens. Tranquillisers do not help remove it and it remains the most common of all psychiatric disorders. Many deaths from alco-holism are rooted in depression; the patient drinks because he is depressed but, after an initial period of elation, the alcohol merely increases the depression.

In 1992 we in the UK donated £82 million to cancer charities, £43 million to animal welfare, but only £6 million were donated to mental health. Depression is treatable by drug therapy, ECT, counselling, psychotherapy and social therapy (via better public attitudes). However, many other illnesses can lead to it – for example, cancer, stroke or diabetes. There are those who believe that mental illness is about moral weakness, malingering or wallowing in self-pity, especially when there are no physical signs. An unsettled childhood may be a root cause, but depression can also be an attempt to shift one's guilt onto someone else, and to undervalue oneself. The patient can often suffer from a lack of someone in whom to confide.

Spike Milligan had an unsettled childhood in India where his parents moved about from place to place. After he was wounded, his officer called him a coward, even though he had served right through the North African campaign and had been mentioned in dispatches for bravery. At the end of the war he had a massive loss of confidence. Yet, he was able to write scripts for the Goon Show at the rate of two per month for six years from 1953 to 1959. All this time, he was fearful of financial failure although he was riding high on hypermania. Loud noises still upset him.

The depressive patient needs to share problems, to learn a more positive approach to life, and to see the illogicality of some of his behaviour. Milligan was too ill for any of this to succeed at first. Some drug therapy reinforced with psychotherapy is used nowadays. Reassurance alone is not sufficient in severe cases. Tricyclic antidepressants (TCAs) have been used since 1956; they can take up to four weeks to show any effect, but some four out of five patients respond to them. Milligan was one of the 20 per cent who did not! Lithium was found to treat mood swings as long ago as 1949 and Spike Milligan has been helped by this. Lithium is not a permanent cure and can take up to a year to work, but it can reduce the severity of the illness. *Depression and how to survive it* also relates how Milligan took TCAs. Tranquillisers can help if there is a danger of suicide, ECT has also helped here. Five per cent of the population suffer from depression but few ever see a psychiatrist for fear of being labelled as insane. The GP finds it difficult to persuade these patients to visit a consultant.

It is important that relatives help the patient to see the better side of things, to keep themselves occupied, to tell others how they feel, and to cry if they feel so inclined. Depressive individuals always feel isolated. Friends should ignore any change in personality, help them stay on their medication, look for any signs of heredity, and do not accept depression as being inevitable. The sufferer should never make rash decisions while depressed, such as resigning from work or abandoning friends or partners. Drinking always makes matters worse. Spike Milligan, although a Roman Catholic, ended his own marriage. He now enjoys the support of his present wife. He survives since having felt better in 1991. He is now determined to continue with his life and he has a clear understanding of his illness.

Harold was one of my first clients and I will never forget visiting him in November 1984, sitting alone in his flat with no heating on. He shivered in his overcoat, staring at the cold electric fire, saying that the meter had been broken into. He had been on a resettlement course but complained that the Disablement Resettlement Officer (DRO) had nothing to offer him. He said he had no friends but, when pressed, admitted that he did attend the church across the road. I wrote to the vicar suggesting that some voluntary work might help Harold back into the community. I also contacted the Regular Forces Employment Association (RFEA), but nothing came of it.

Harold had served in the RAF and became a Corporal working on the loading of transport aircraft. He had only done odd jobs since coming out of the Service; never qualified for a War Pension – he simply felt too gloomy to apply. He was in his late forties, physically fit, smart in appearance, but clearly carrying a great burden of guilt for some reason or other. His GP prescribed him tranquillisers.

At Christmas 1984, I sent Harold a cheque for £10 as a present. And when it was not cashed, I called in February 1985. I found he was living in another flat nearby with a very nice lady friend who was obviously sensible and caring. Janice was clearly not pleased with the possible loss of the £10! She took the cheque out of Harold's wallet saying that she would cash it. I also gave him a cheque, payable to the Electricity Board to cover his arrears. Janice took charge of this also. He seemed to be in some sort of a daze which I attributed to his medication. Harold had met Janice at the local church and she was a bustling sort of person who worked at the hospital.

They were married soon afterwards and they told me how relatives they had not seen for years turned up at the wedding. This was the happiest time I ever saw Harold and, soon afterwards, he went to Tyrwhitt House. I next saw him in hospital and he told me he was free of headaches, having recovered from a brain haemorrhage. There were, however, some ominous signs in his complaints about Janice's spending on frivolities. He refused to let me help, saying that this would only encourage her to be even more extravagant.

Their marriage only lasted three years and Harold was depressed every time I saw him from then on. He was unable to find work and felt insecure; meanwhile Janice was doing well, but spending the money she earned. He felt trapped and admitted to losing his temper. The doctor put him on an indefinite sick note. For most of 1988 he hardly went anywhere.

A year later, they had split up and divorce proceedings were begun. I wrote to the Council to get him rehoused, and he was moved from poor lodgings to better accommodation. The next time I saw Harold he was in a flat, in a tower block, in November, with no heating on. The wheel had gone full circle. There was a community centre, and there was the church, but he would not go out. He said he was too frightened to see his psychiatrist, for fear of what he might say. The CPN called, but that was all. Harold said he could not cope with Tyrwhitt House, although I offered to take him there personally.

In 1991, he went to a tribunal and was awarded a 20 per cent War Pension for mental illness aggravated by the Service. Harold began to feel better, and we had him admitted to Kingswood Grange, the Society's permanent home for veterans; the vicar transported him there. However, after only a few days, he discharged himself, and I next saw him back at the flat, as gloomy as ever. He occasionally saw Janice, but she was finding it hard to remain friends with him.

Two years later on, I attended Harold's funeral. In a fit of depression, he had taken all his tablets at once, and the hospital staff were unable to save his life. The church was full, with representatives from the hospital, the RAF Association, relatives I had never seen before, parishioners, and of course, Janice. The Vicar said, 'This is a sad day for which I know everyone here must feel some responsibility'. How right he was. When it was too late, we could all have said that we ought to have done more for poor Harold. Some of us had tried, but he put up the shutters on practically every occasion.

Towards the end, it had not been a case of Harold being short of money. Ironically, his pension had been increased, and there had been some arrears awarded to him as a lump sum. Yet he never had a holiday, and I imagine the family continued to stay away. The Vicar was right in that we were all, to some degree, responsible for Harold's fate. Yet, one of the aspects of this illness is that the sufferer transfers his guilty feelings to other people. In this respect he had succeeded in involving us all.

Andrew joined the RAF Regiment in 1964. He had to be medically evacuated from Cyprus when he could not sleep and suffered from nightmares. After working as a civilian driver, he became depressed and had a nervous breakdown. He could not cope with life and felt inadequate. Andrew then did labouring jobs. When I met him he had been divorced twice, and was living in a Council house with a third wife. He had periods of depression interspersed with work as a taxi driver. He went on to develop agoraphobia and did not leave the house for weeks on end. Whenever he felt better, he read books about the RAF and made model aircraft. Carol contacted the Society, after picking up the address from leaflets in the public library.

Andrew was admitted to Tyrwhitt House for a successful stay, and it was decided to promote a case for a War Pension as well as provide a holiday in Bournemouth, from which Carol and the two children benefitted enormously. They had all been cooped up in their very small house with Andrew, a recluse, for far too long. SSAFA paid for the children and the Not Forgotten Association paid for Carol and Andrew to spend a week at the Cedarwood Guest House. The visit was photographed by another client, and the pictures appeared in the Society's magazine in 1991.

On being turned down for a pension, although he received a lump sum for tinnitus in his ears, his condition began to worsen, although further visits to Tyrwhitt House did help. He was quite unable to work by now, and Andrew took the death of his father very badly.

The next piece of news which broke was that Andrew had left Carol and the children and had gone to live with a lady who was nearer to his own age. She at first gave him a job as a driver. He only lasted four weeks in this work, but surprisingly he and Margaret stayed together. Although the break-up with Carol upset him sufficiently to make him suicidal, Andrew had found that he really could not cope with looking after the children.

Margaret and Andrew continued to be happy together

although he never worked again, so she took him to work with her, as company. A domicilary visit by members of the local War Pensions Committee led to a 50 per cent pension for personality disorder, together with the award of the Unemployability Supplement.

There were arrears of council tax, he lost his hearing aid, and broke his spectacles. Depression was never far away. He continued to visit Tyrwhitt House and, on the whole, with support from the Society, life with Margaret was pleasant. They were married just before I retired. Much time was devoted to Andrew and his problems, but I wonder just how many of us realised that it was unipolar manic depression which lay at the root of them.

Helen had served in the WRNS during the Second World War. Although in her seventies, I found her cheerful, living alone. Her problems were that she had become agoraphobic, and clearly had had no contact with her relations after the loss of her husband thirty years previously. Helen was diabetic and struggled to keep herself off most of the drugs she had been prescribed. She continued to take an anti-depressant.

Helen was known to the War Pensions Welfare Service as she had a 20 per cent pension for personality change aggravated by War Service. She had refused all other forms of help except, being keen on handicrafts, she welcomed periodic visits from the War Pensions Handicrafts Officer. I was quite unable to persuade Helen to come to Tyrwhitt House; all kinds of problems were raised, especially the male environment there, but it was Helen's depression which prevented her from mixing with others. I imagine that other callers were probably deterred from making further visits.

I only had a few clients from the ethnic minority groups. There never were very many of them in the British Forces, anyway. **Dennis** was born in Jamaica and came to England as a child. He attended secondary school to eighteen and achieved his GCE 'O' levels of which he was very proud. Dennis served as an infantry soldier in Northern Ireland, Belize and Gibraltar. There were never any problems connected with his background, I was informed, and the Army invited him to extend his Service. But he declined in order to settle down as a clerk in Local Government.

All went well for a few years but Dennis began to have increasing periods of absence caused by bad mood swings; eventually, the Authority had to discharge him. He felt that Service in Northern Ireland had had nothing to do with this, and the situation conveyed to me showed all the signs of a depressive psychosis. The local psychiatric hospital had been treating him as an in-patient and, later, as an out-patient.

Dennis was able to benefit from Tyrwhitt House, and the hospital transported him to us at first. They eventually found him a sheltered flat with some other patients. He did make some efforts to apply for jobs and he attended the hospital day centre, as well as local evening classes. When I last saw Dennis, I felt that his problems were two-fold; his association with a mental hospital, and his ethnic background. This was a great pity because Dennis was physically fit and well educated, but I did detect in him a cynical view of life. The hospital eventually discharged him and employed him in their workshops, for which he was paid a nominal wage. At least this was something, and he began to attend Tyrwhitt House on public transport. This in itself was a sign of some improvement.

Angela joined the RAF at eighteen. She was a driver, but became interested in parachuting and sky-diving, becoming very proficient in the sport. She took part in international events until she was seized with a panic attack whilst overseas competing, and depression set in. For the next six months, Angela was an in-patient at an RAF hospital. She received a 20 per cent War Pension for a condition aggravated by the Service. At least this was a help. It had appeared that the sky-diving had triggered off something which had lain dormant throughout her early days.

Angela was diagnosed as manic depressive and was out of work for the next two years, until a job came along which she was able to hold for the next four years. A further period of depression ended her marriage, by mutual consent. Like Spike Milligan, Angela gained an insight into her condition, realising that, when overtaken by bad moods, it was best to lie low and live alone. When I first saw her she was on a mixture of strong medication and said that, at times, she could do nothing at all.

Angela came to Tyrwhitt House and we helped her get an allowance for a lowered standard of occupation. This helped her financially. She did, however, keep in touch with her parents. When I last saw her, we were trying to have her War Pension percentage reviewed, although I was careful not to build up her hopes too high. The next stage might be to persuade War Pensions to make her award 'attributable' to the Service. After all, they had provided the facilities and encouraged Angela to take part in events which turned out to be too much for her.

Neuroses are described as less serious psychiatric disorders in which some insight is retained, although the patient cannot help his or her behaviour. It is estimated that a quarter of all GPs' patients have some degree of neurosis. In fact 10 per cent of the population suffers from a neurosis of some sort. Generalised anxiety was a feature of many of my clients, they worried about trivia; but the more striking problems came with the serious fears such as agoraphobia. Then there were the obsessional neuroses, and states of depersonalisation. These are just a few of the types of conditions I actually saw for myself. Their causes might be generic, or environmental, or triggered off by events such as active Service in the Forces. Into this latter category must fall PTSD.

Post Traumatic Stress Disorder has been the most rapidly recognised psychiatric syndrome of recent years, yet it has always been there, in historical events and in literature. By the end of the Second World War, half a million American infantry soldiers had been taken out of the front line suffering from battle shock. Shakespeare depicts how the character of Macbeth deteriorates after the killing of Duncan, and there is a similar development in the character of Hamlet. Tragedy on a large scale produced PTSD, recorded as 'melancholy madness' after the Great Plague of London in 1665, by Daniel Defoe and Samuel Pepys.

Vietnam recently brought matters to a head when it was found that one-third of those who had served there became abnormally stressed. In recent years it has been estimated that there are still almost half a million ex-soldiers with PTSD. What was worse, many of those who returned from Vietnam were not regarded as heroes, but were shunned as criminals. In Vietnam, units were continually changing, so very few of the men made friends there. This should be compared with the events of the First and Second World Wars where the comradeship of the British soldiers helped see many of them through the worst of their anxieties. Some

Falklands veterans came home with a 'macho' idea that grown men do not cry but bear their grief in silence.

PTSD[1] can follow a distressing event which is far outside the normal range of human expectation. It can be brought on by a serious threat to one's life or body, a serious threat to one's family, the sudden destruction of one's home or community (as in Bosnia), seeing another person being killed by an accident or by some physical violence, or learning about a serious threat to a close friend. The most outstanding symptom is the one that relives the event and won't go away. The victim relives sights, sounds or even smells. A 'reminder' incident can start the process off all over again.

Victims go to great lengths to avoid such reminders, they can become emotionally numb, feel detached, or become artificially 'macho'. They cannot sleep, lack concentration, are suddenly startled, and become aggressive with outbursts of frustration and anger. On the other hand, the anger may be bottled up inside. Anniversaries or similar events will start the feelings off again. There can be a lack of interest in anything and the future will look gloomy. Sometimes reactions are delayed for months or even years, as with some Second World War veterans. Individuals can cover things up for years, seeing no good in reliving bad memories. They won't admit to anything wrong for fear of losing their job or being labelled as insane. Some think, wrongly, that it will all go away. Victims avoid reminders, some resort to drugs or alcohol, but diversions do not last. Physical ailments can arise, and there can be a fear of crowds of people.

Accidents can lead to PTSD in the witnesses, the rescue Service personnel and medical staff, as well as the victims themselves. Policemen and firemen can feel no longer able to do their work, having been overwhelmed by events. A family history of depression or anxiety can render the subject more likely to suffer from PTSD, but even the strongest characters can succumb. All victims are better able to cope, with the support of their families, rather than in isolation or in an atmosphere of hostility.

After the disastrous fire in the football stadium at Bradford, a leaflet was issued to survivors. It warned them of the feelings to expect: fear of injuries to oneself, fear of being bereaved, or of breaking down mentally, fear of another crisis happening, helplessness, yearning for what has been lost, guilt and regrets for things not done before the event, shame for not having done more, anger at the shame and towards those who caused the disaster, and anger at any lack of understanding on the part of others.

In such disasters feelings have been intensified where many died, or if bodies are not discovered, or if the relationship with the deceased had been dependent, or unhappy and unresolved. Survivors or those involved often feel exhausted and drained of energy. They may also suffer from nightmares, memory loss, dizziness, shaking, nausea, diarrhoea, loss of sex drive, quarreling with others, and recourse to alcohol and drugs. The advice given at Bradford was to try not to repress one's feelings, to help the others, to go through the event and confront it, to accept support but try to enjoy some privacy, to share one's feelings but try to lead a normal life, to drive carefully and avoid accidents in the home. Victim Support is an organisation set up in the 1980s: they help victims of burglary and violent crime, and are listed in Appendix C.

The Trauma Trap by Doctor David Muss, published in 1991, refers in detail to what he describes as 'neurolinguistic programming' (NLP). This is a process which helps control one's recall of a tragic event, either as a victim or as an observer. There is an impact of events scale (IES) questionnaire to be completed before and after treatment. The treatment recommended is what is known as the rewind process.

The patient relaxes his muscles from the feet upwards, tries to float out of himself and tries to remember the event as if he were an observer. He then imagines himself in the projection box of a cinema, watching himself seeing a film of the disaster. He sees it forwards and backwards until he sees himself start from just before the event, and this should be a happy scene.

During the rewind, he imagines he is in the film and lets it run back to the happy ending. Whenever the event upsets him he should be able to return to the happy scene of just before the disaster. He should not avoid the most horrific parts. Anyone can do the same for other bad memories, and this technique works for observers or for persons who have to identify a body. The advice given is to get into the scene and then rewind to the happy ending which is part of the event. One should practise the rewind until

IES QUESTIONNAIRE

NAME————————————————————————————————————

ON—————————YOU EXPERIENCED————————————————————

(DATE)   (LIFE EVENT)

Below is a list of comments made by people after stressful life events. Please check each item, indicating how frequently these comments were true for you DURING THE PAST SEVEN DAYS. If they did not occur during that time, please mark the 'not at all' column.

FREQUENCY

	Not at all	Rarely	Sometimes	Often

1   I thought about it
    when I didn't mean to

2   I avoided letting myself
    get upset when I thought
    about it or was reminded
    of it.

3   I tried to remove it from
    my memory.

4   I had trouble falling asleep
    or staying asleep, because
    of pictures or thoughts about
    it that came into my mind.

5   I had waves of strong feelings
    about it.

6   I had dreams about it.

7   I stayed away from reminders
    of it.

8   I felt as if it hadn't happened
    or it wasn't real.

9   I tried not to talk about it.

10 Pictures about it popped into
    my mind.

11 Other things kept making me
    think about it.

12 I was aware that I still had a
    lot of feelings about it, but I
    didn't deal with them.

13 I tried not to think about it.

14 Any reminder brought back
    feelings about it.

15 My feelings about it were kind
    of numb.

you get it right, and then fill out a second IES questionnaire.

Once out of PTSD the patient can attend to the side-effects affecting physical health. In some cases, a course of treatment for depression for a couple of months can help before the rewind process is attempted. PTSD is now accepted in law as a reason for claims for damages, but there is still a general feeling that physical injuries are more disabling than mental ones. As was felt by many of the war veterans, there is an attitude that compensation constitutes a public recognition of one's loss and, in a sense, it can give the victim back his self-respect and control of his own life. This can help him return to normality. This was certainly the case with my successful War Pensions claimants. It is now possible to experience mental stress without being regarded as insane, and this is an important step forward for all our clients. The highest payment made to a victim of the *Herald of Free Enterprise* ferry disaster was £30,000 to a man of fifty-four for depressive illness, pathological grief, and severe PTSD.

Many people have had to give up their jobs because of reminders of an accident and its memories. Employers ought to see that it is important to try to get their staff back on duty if at all possible; but this may take some time, which ought to be allowed for. Some American firms employ a counsellor to deal with industrial accidents and crises; by this means, there is less likelihood of being sued for damages or of absenteeism. Most employees on the *Herald* left P&O altogether, while men from the Piper Alpha disaster were unable to return to working on oil rigs. Debriefing after an accident is very important and can be a requirement under the Health and Safety at Work legislation.

If help is not received, the trauma will take over and rule the individual's life as sympathy from others dries up. The victim can feel guilty for being difficult and, if he strikes his wife, she can, with reason, end the marriage. Children can become aggressive if they see their parents quarrelling. Thirty-eight per cent of Vietnam veterans' marriages broke up within six months after they returned home. The British Psychological Society advises against counsellors becoming too personally involved with their clients. The Society publishes a book entitled *Psychological Aspects of Disaster* (Sept. 1990, available from St. Andrew's House, 48 Princess Road, East Leicester LE1 7DR). In this book, the rewind technique is recommended to spare the counsellor the full details of all the events. The counsellor should attend to other problems and act without delay before clients drop out, for it is often the case that clients have become disillusioned by other methods. Individuals seeking compensation may have no wish to recover. It has been estimated that a third of those wounded in battle can have PTSD. They will have probably been told they were immature and should pull themselves together. The rewind process is a means whereby the individual can try to pull himself together.

*   *   *

Trevor served in the Parachute Regiment from 1976 to 1990. He purchased his discharge after three tours of Northern Ireland and Service right through the Falklands campaign. In Ulster, he had been acting as an undercover agent and, when I saw him, although he had been out for three years, he still experienced flashbacks to some of the extreme dangers he had survived. Trevor had had a number of jobs but the trouble was that his wife, for whom he had left the Army, did not want to know about his problems. While I talked with him in the sitting room, she watched TV and never said a single word to us. I did, however, make some progress towards giving Trevor an insight into his problems.

We were able to place Trevor on a stress management course at Royal Naval Hospital, Haslar. When I last saw him, his wife was coming to collect Trevor and take him home for Christmas 1993.

The Airborne Forces Association helped financially, and the idea of applying for a War Pension, which had never occurred to him was taken up. Trevor was to receive a 20 per cent pension for PTSD. His employers, where he worked as a security guard, promoted Trevor to Security Manager and, when he finally left his wife, problems with the Child Support Agency were sorted out. He now attends regimental activities and stays at Tyrwhitt House from time to time. This is a recent story of the Society successfully supporting a deserving client.

Reg was an older man who had served in the Fleet Air Arm in the Second World War. He was on an aircraft carrier torpedoed in the Atlantic, with tremendous loss of life, as the great vessel turned over on its side before it sank. He remembers the cries of drowning men whom he was powerless to help and this was to haunt Reg for the next forty years.

The television pictures of the *Herald of Free Enterprise* which capsized off Zeebrugge brought it all back to Reg whose GP advised him to apply for a War Pension. This he did with the aid of the Royal British Legion. Reg was notified to me after his wife committed suicide in 1993. She had nursed him through all his anxieties, but she was losing her sight and could not bear the thought of herself becoming dependent on anyone. We took Reg into Tyrwhitt House where he responded well to the comradeship he found there. His 30 per cent pension for PTSD was increased to 100 per cent and Reg was delighted when he was awarded a lump sum for arrears of payment, which amounted to several thousand pounds.

Generalised anxiety is an aspect of neurosis, probably hereditary, but which can be made worse by events. I have discussed PTSD but this other aspect of nervous anxiety can be brought about by sorrow, guilt or disgrace, feelings which may build up; as the involuntary side of the nervous system takes over, fear

Tyrwhitt House.

begets more fear. This brings about physical symptoms such as a churning of the stomach or a flash of panic. The basic advice given is not to fight bad thoughts but to accept that they do not matter. The patient needs support along with freedom to relax. He also needs to feel that these sensations are not normal and will eventually pass.

Individuals can withstand stress so long as the body receives the chemicals it needs to make the hormones the brain requires to combat stress. When stress builds up, the supply of hormones runs down and more fears develop. This is a simplistic view but, with help from an understanding GP and friends, the system works. We are in effect healing a tired mind. 'The Latest Help for Your Nerves' by Doctor Claire Weekes (Angus & Robertson 1989) follows this line of thought in much greater detail.

Another useful book has been *Life Without Tranquillisers* by Vernon Coleman (Piaticus 1985). The world's greatest addiction problem has been that of the benzodiazepines, Valium, Librium, Mogadon, and others. One person in ten has taken them. They work for a short time after which increased dosage is required, and a spiral of addiction sets in. The human body was designed to face stress by activity, by fighting. The body becomes alerted, the adrenalin surges, muscles tighten and heartbeat is increased. Today we face stress caused by excessive physical or mental

Occupational therapy at Tyrwhitt House.

A Wide Range of Cases

activity, or by sudden shock in battle or elsewhere. There are the stresses of today's competitive world and the thresholds of human capacity to endure stress can vary considerably. Break-up of family life in the 1950s and 1960s produced stress on a large scale. Millions believed that the NHS entitled them to a healthy life, that mental anxiety could be treated by taking a few tablets, all on prescription, and that there was no need to put up with nervous anxiety. We had all paid in advance for our tablets, it was thought.

The benzodiazepine drugs were discovered in 1958, and a new chemical substance produced was named chlordiazepoxide, marketed as Librium in 1960. One wonders whether there is any connection between the name Librium and the Latin word for 'freedom' or 'liberty'. Librium can be absorbed quickly and it changes the rhythm of the chemical activity in the brain, thus inducing drowsiness and relieving tension and anxiety. Amounts of this drug in the body can accumulate. Variations of this drug were produced and diazepam was sold under the name of Valium; it was more potent than Librium. Taking them became fashionable, a cure for insomnia, or a panacea for all ills. Medical schools have treated pharmacology like psychiatry, as a fringe subject; therefore, GPs were influenced by drug company representatives and their brochures. The real problem was that benzodiazepines did not actually cure anything; the patient felt better temporarily but became addicted to the medication.

By 1979 the World Health Organisation (WHO) estimated that there were a thousand psychotropic drugs, with about seven hundred Valium-like substances on the market. Hospitals prescribed them in order to give the staff some peace at night. They can make the elderly even more confused, make driving risky, and they do not mix with alcohol, anaesthetics or caffeine. Their side-effects include low blood pressure, skin irritation and dizziness. Addicts can accept these conditions and become zombies. I have visited clients in just such a dreadful state.

The only cure is to be weaned off the tablets by halving the dosage every two weeks. The patient is advised to try relaxing the mind by day-dreaming, imagining a quiet scene, lying down and drawing the curtains. He should try to write down his problems and their solutions in a note book, then plan events ahead. When able, the sufferer should let off steam and mix with brighter people for stimulation of the mind.

Neuroses can produce clear-cut types of nervous conditions, but there is also the state of all-pervasive or generalised anxiety where the affected adult cannot identify any specific fears. He has them all the time, and becomes hypersensitive to stress.

* * *

Alex was one of six children. He had to go to work as a clerk until he joined the RAF at eighteen. He went into Air Traffic Control, serving in Egypt, Sudan, Cyprus and Malta, gaining rapid promotion in the 1950s. As an Acting Flight Sergeant he was in charge of fire-fighting on an airfield. However, when he was posted, yet again, this time to Lebanon, he developed nervous anxiety. He was invalided out with generalised anxiety syndrome and for this Alec received a War Pension.

Alex's marriage did not last, and I found him living alone, going through periods of great stress interspersed with near normality. When Alex came to Tyrwhitt House for the first time, he did not stay the full period allocated to him because he was not accustomed to living with others. On a subsequent visit, he stayed the full time. The council modernised Alex's flat, and when I saw him for the last time, he was doing well.

Nervous anxiety can crystallise into certain more specific psychosomatic conditions. I was to meet agoraphobia, claustrophobia, obsessional compulsive disorder (OCD), and depersonalisation in my clients. In chapter 3, I mentioned Harry Byrne who suffered from battle fatigue in Italy, and who, many years later, became confined to his own home for years on end. Then, in chapter 8 we met Billy Douglas, a victim of war in Northern Ireland. Billy was to recover from agoraphobia with the help of a caring wife and the good fortune to find employment in the open air, working alone. When Billy's wife became ill he was able to support her.

Ronald lived alone in an upstairs flat, refusing to venture out beyond the corner shop. His was an agoraphobic based on an obsession that his landlord, who lived below him, was trying to get into his flat in order to discover reasons for evicting him. I found Ronald's flat was spotlessly clean with everything in its place. He said he had felt like this for ten years and he only went to the shop when the owner himself went out; he maintained a constant vigil on the flat below. Ronald was convinced that the landlord was trying to steal from him. When I offered to pay for a new lock on the front door, Ronald quickly changed the subject. The flat was in need of redecoration and I suggested we could pay towards having it done; again there was no response. Ronald had served in the Army throughout the Second World War and had a 20 per cent pension for psychoneurosis, but my attempts to have him reassessed by a doctor fell on deaf ears. I never succeeded in doing anything for Ronald because he never co-operated.

Another obsession is claustrophobia and in chapter 4 we mentioned Bill, a stoker in the Royal Navy who had been torpedoed several times. Understandably he had a fear of being shut in. We helped him keep his ancient car on the road as a means of escaping from the confines of his house, all the year round.

In some cases of stress, the brain switches itself off and dissociates itself from the world outside. One of my clients had a wife, Elsie, herself ex-ATS, who suffered from dissociative state. **Fred** was on a high percentage War Pension for nervous anxiety but, having taken on all the housework, I would say that he had no time in which to worry about himself. There were the occasional mental blackouts for which Fred would go into hospital while relatives took Elsie in for a few days. This worked so long as it only was for a few days as none of the family said they could manage their mother for very long.

Whenever I visited them, Fred would be cheerfully cleaning the house or doing the washing, often drying it indoors. The house was very untidy but what else could one expect? Elsie spent all the time staring at the wall. When I got to know her, I was able to exchange the odd word but that was all. We eventually found a solution by which Fred came to Tyrwhitt House for up to four weeks while the GP placed Elsie in a geriatric hospital. There she could be properly looked after for a while. Fred said he was sometimes too tired to sleep; this was probably because he had to think for both of them, including giving Elsie her medication, dressing her and taking her to the toilet. Fred still comes to Tyrwhitt House and he benefits from a change of scene and a rest from his responsibilities. The doctor appeared to assume that Elsie was a patient for whom nothing could be done beyond giving her tranquillisers. Clearly, it would have been wrong to separate them permanently.

Obsessional compulsive neuroses refer to a repetitive thought or idea which can dominate one's whole life or existence. Sufferers may be trapped by senseless rituals such as washing, checking and petty details. Many keep their obsessions a closely-guarded secret, and the fantasy is that unless the victim obeys these compulsions, something dreadful will happen. This illness can be confused with schizophrenia because the end result, as when voices are heard, can be similar. Medication does not always seem to be effective. A better insight might be the best road towards improvement. Some become obsessed with the detail of their work, continually checking everything, and this can lead to their downfall. This was to befall two of my clients who became perfectionists – always right, workaholic and lacking any consideration for others.

Matthew lived alone. He had a 20 per cent War Pension for obsessional neurosis. When I met him, he had been suicidal. His pension had only been awarded to him after much effort, which had left him even more anxious. As Matthew had been invalided out of the Army after ten years' Service he only had a small Army pension giving him a total income of under £50 a week. Because of his small pension Matthew alleged he could not claim Supplementary Benefit or Unemployment Benefit. I was able to help him by providing some shoes and clothing.

Matthew had begun to be obsessed with the details of his job as a hospital administrator in the Far East. He was a Sergeant in the RAMC, and his repeated checking on everything drove everyone else to despair, especially when Matthew began to be forgetful. He was put on phenobarbitone and posted to a military hospital in the UK where he began to behave in exactly the same manner as before. On leaving the Army, periods of employment were interspersed with hospital admissions. Whenever he found work, he was usually dismissed for the reason that he achieved very little and upset large numbers of others by complaining about them. He was always perfect!

He had been married twice before; each relationship had failed and I imagine his partners never understood Matthew's problems. His War Pensions Tribunal had been conducted against an assumption that Matthew's work was similar to working for the NHS and therefore the Army could not be blamed for his illness, which would have occurred anyway. The argument which prevailed was that being a military Wardmaster was more stressful than a civilian. This was because Matthew had to comply with military discipline in satisfying his superiors and working with his subordinates; moreover, being overseas, more responsibility was placed on him, as a Sergeant, than would have been the case in a civilian hospital, at home.

Matthew was befriended by Elaine, an older lady, and a kindly, caring person. They were married and the Society helped towards the cost of materials to enable them to redecorate their flat. I tried, without success, to get Matthew back to work. The RFEA were friendly towards him, but the reports about him not fitting in with other people weighed against him. The DRO was also approached. Meanwhile, Matthew and Elaine were happy together. They showed me their wedding photographs with great pride, and Tyrwhitt House was not mentioned. Twelve months later, Matthew's situation had changed dramatically for the worse. The marriage had not worked out, although he and Elaine were still friends. He had been to Tyrwhitt House but only stayed there three days. I gave him £10 towards a new Calor Gas cylinder of which he was in desperate need. From then onwards things could only improve. We took him back to Tyrwhitt House and the Society helped him claim a higher War Pension percentage. He began to co-operate with others and this was a promising sign. However, when I last saw Matthew, I do not think he was taking any medication; this was bad news in that any chance of him finding work without this kind of support was going to be very difficult.

Donald was a Sergeant in the RAF. Like Matthew, he was employed in administrative work, but he never qualified for a disability pension as his duties were never regarded as unusually stressful. Donald had been in clerical work but eventually transferred to General Duties Branch where he was Sergeant in charge of discipline at an Air Force station in the UK. However, Donald became obsessed with all the trivia of RAF routine and procedure, driving everyone else into a frenzy of frustration. Not surprisingly, his wife left him to fill out his RAF forms and polish his boots on his own. In order to retrieve his marriage, Donald left the RAF, after twelve years, but his Service pension, under the Regulations, was frozen and could not be drawn until he was sixty.

Donald became a Traffic Warden and he threw himself into enforcing the rules just as he had done whilst in the Air Force. He became not only unpopular with the motorists, but also with his own colleagues. The unkindest practical joke of all was when someone took Donald's own car and parked it on the yellow lines outside the Police Station. The Chief Constable was not amused, and Donald had a nervous breakdown. His obsessional behaviour had rendered him unemployable.

This time, his wife stood by him and she obtained part-time work to help them just manage financially. I tried all kinds of ways to help Donald find work but nothing came of it. The DRO admitted to me that he would have stood a better chance had he been physically disabled. We took him into Tyrwhitt House but his case for a War Pension was never accepted. The Society was, however, able to help them considerably when our Medical Officer showed the Service authorities that Donald was unlikely ever to work again because of his illness. As a result of this, they authorised immediate payment of the Service pension he had already earned.

Donald settled down considerably after this, and all we now do is to pay an occasional visit to see that all is well. His wife has become more friendly and they spend some of their time caring for their grandchildren, when they come to visit.

*   *   *

A different category of mental illness is that of the personality disorders. These include those conditions which the Service never knowingly accepted, such as immaturity and inadequacy. There was Alec, mentioned in chapter 5, who was looked after by his brother, having been discharged from the Army, after only a few months.

Then there is psychopathy and criminal delinquency. I never visited the secure hospital at Broadmoor, although it was inside my area. I did sometimes receive letters from the inmates asking me to help them escape! Two other disorders are addiction and deviancy. I met several ex-Servicemen who had succumbed to alcohol. Some of them died penniless, in squalor, and lacking the will to reform. Because of their condition, they lacked the friends who might have been able to help them through the unpleasant stages of withdrawal. Drug addicts were not encouraged by the Society, for fear of them passing on their addiction to others.

When I have related how I tried to help **Edwin**, at his request, and what happened, you may understand why the charities generally, and the NHS, simply cannot devote much by way of time and scarce resources to cases of addiction.

I found Edwin in a poor lodging house in one of the cities where I worked. He was on treatment for alcoholism but the therapy prescribed, that of a drug called Heminevrin, was itself addictive. The poor man lay on the floor with an empty gin bottle in one hand and a jar of pills in the other, not knowing which he craved most – the drink or the drug. Here was a man in his forties, lonely because of his addiction, needing support if he was to salvage his life. He had reached rock bottom. There are half a million alcoholics in the UK; drugs are useful for masking withdrawal symptoms and they can act as a deterrent or reduce the desire for drink in ex-alcoholics. However, the danger of returning to drink will always be present. The heminevrin was prescribed by his GP and, although his speech was slurred, conversation was sensible. Edwin showed me the craft certificates he had earned years ago as an apprentice painter and decorator. He admitted that, without the drugs, he had no will to live. His lodgings, described as a 'stop off', had been found by his daughter from the local newspaper adverts, as she could not have him to live with her in his drunken state. The owner said that Edwin gave them no trouble. He had a small War Pension for tuberculosis contracted whilst in the Navy.

I visited Royal Naval Hospital, Haslar where, on my recommendation, they agreed to take Edwin into their detoxification unit just as soon as the necessary medical reports could be obtained. I referred the case to the Society's Chief Consultant and it was decided to take him into Tyrwhitt House direct from Haslar. The idea was that he would later go to Milner House, the Society's hostel, where permanent accommodation, with some work and supervision, would be available.

On the appointed day, I collected Edwin from his lodgings. He was wearing his best suit, and looked very smart. He had not had a drink for some time, he told me. However, I took the sensible precaution of receiving an assurance from the proprietor that they would take Edwin back into his lodgings, if it was ever necessary.

All went well. The staff at Haslar weaned him off the drink and the tablets and, by Christmas he was ready for discharge. I personally took him to Tyrwhitt House, taking care not to stop at any public houses on the way! Edwin had a good Christmas, he enjoyed his food, was visited by his family, and he did not drink.

In the New Year, he went a mile down the road, to Milner House. Again, all went well for a couple of months until the Warden rang me to say that Edwin had disappeared and that the Home's drugs cabinet had been broken into. Edwin returned, a changed person. Now he was bombastic and he boasted to the others about what he had done. Namely that he had gone back home on the train, called on his old GP, hoodwinked him into believing that he was back in his old lodgings, and obtained a prescription for a large quantity of heminevrin, most of which he had consumed. He had to be packed off on the next train, after I had checked with his old lodgings that they would take Edwin back.

We all felt let down by this patient. I do not think that any of us could have done more for him. Work had been waiting for him to start at Milner and materials had been assembled for him to use. He admitted that he had been well treated at Haslar, where I had seen him shake hands with all the staff on the day I collected him.

There is a postscript. Many months later I was asked to call on his daughter, for whom I felt very sorry indeed. She lived not far from where Edwin was and she pleaded with me for the Society to have her father back. The poor woman was afraid that he would turn up drunk and frighten her small children. I informed the GP of this situation. He, in turn, in the opinion of our Consultant, had behaved unprofessionally in supplying the drugs after Edwin had been treated for his illness. That doctor should have co-operated with the officers of the Society, in the best interests of our patient.

I was to encounter other alcoholics, some of them quite harmless to other people, but I never again attempted to reform anyone. **Eddie** lived in squalor and was addicted to barley wine. His only friend was the publican who served him. Periodically, Eddie would take an overdose of tablets and have to be rushed to the local hospital. He always denied that he was an alcoholic and claimed he could stop drinking whenever he liked. However, he always declined to consider Tyrwhitt House where, I said, he would not be encouraged to drink. Because of physical problems he drew the Mobility Allowance and he had a small War Pension, all of which, I imagine, went on barley wine. Everyone involved with Eddie was aware of his problems – the GP, the Social Worker and the War Pensions Welfare Officer. Sadly, one day Eddie took one overdose too many.

Another personality disorder is sexual deviancy. The Royal British Legion asked me to visit a client who had a small War Pension for anxiety neurosis. **James** had a good case for an increase, but he refused to attend a Tribunal which had been arranged locally. James lived in a tower block in the Midlands, a youngish looking man, ex-RAF. He alleged he knew when crimes were about to be committed before they happened. He rarely went out of his seventh-floor flat except to shop nearby, and a few extra pounds on his pension would have helped.

My suspicions were aroused when I saw, on the walls of his flat, pictures of a glamorous looking woman with blonde hair and a well proportioned figure. I made some remark, and he said, 'Oh! They're me. I used to be on the fringe of show business.' I asked, 'Were you a hostess in a night club?' 'No,' he replied, 'I was a projectionist with the RAF Cinema Corporation!' I made some joke about him showing 'blue' movies, which he did not appear to understand.

The tribunal was bizarre, to say the least. James appeared, wearing his blonde wig and a ladies' two-piece suit. The Chairman opened by saying, 'Good morning Mr-er-Miss-er . . .' James stepped in. 'Just call me Sister Barbara!' I was there to see fair play and James had to be medically examined, on the couch, and we all saw evidence that he was indeed a man. He disclosed, in reply to a doctor's question, that he was on hormone tablets to develop his bust. I think it was his strange behaviour which won his case, and an increased pension for mental instability was awarded.

When I last saw James, he was on a 50 per cent War Pension but, because of this, his Social Security benefits had been cut, and I felt obliged to give him some money towards his arrears of electricity bill payments. He believed he was still being chased by gangsters from the London underworld; I suspect he acquired his stories from reading the papers. James had already sold some of his furniture and said he was leaving for Scotland. I alerted our Scottish office but I never saw 'Sister Barbara' again.

*   *   *

We finally come to the organic states of the mind. The most common are the dementias. It was probably after the Second World War that the Society, originally set up for First World War victims exclusively, decided to offer help to any ex-Service person suffering from any form of mental illness. As a result, about half my clients were war pensioners while the rest, including some of the most deserving, were not. A further complication has arisen with the realisation that the NHS and the Welfare State will not be able to cope with all the problems of the entire population from cradle to grave, for ever. A few statistics will help explain the size of the dilemma the country faces, on account of our ageing population.

Dementia is a generic term for a condition in which the arteries of the brain become diseased. People live longer. There are now over three thousand centenarians in this country, and Britain is one of the world leaders in terms of longevity. Fortunately, the majority are mentally well, all their days. One-tenth of the eight million over sixty-five have some dementia or confusion, and there are a further 800,000 who are depressed and who would benefit from some treatment. Only 17 per cent of elderly men live alone, most live with their wives. Only 6 per cent of the elderly live in homes. In fact we have always had 'Care in the Community', the elderly prefer it this way. One in eight of the elderly at home are housebound. Dementia is the progressive decline of one's mental faculties and, once it starts, there is no recovering from it. It can also occur in the young, when parts of the brain cease to function.

Alzheimer's disease can be an alternative name for senile dementia, and there is a form of it called pre-senile dementia, which can start in middle age. Alzheimer's is named after the doctor who first described the condition, a hundred years ago;

similarly with Pick's disease, Jacob Creutzfeld disease and Huntington's chorea, of which more will be said.

In Alzheimer's, the outer layers of the brain become biochemically changed, and this affects the nerve cells which pass messages to other cells which control memory and learning. Research is continuing into the cause of this. Multi-infarct dementia is a variation of this condition, and this is caused by blockages in the minute cerebral arteries, thus leading to strokes. Some stroke victims may recover. Dementia is not caused by mental inactivity, head injury, or sudden shock; these have been commonly held to be causes, but this is not now regarded to be the case. Jacob Creutzfeld disease, which can affect the young, may be carried by a virus. Dementia is not caused by a virus, nor is it inherited. Sufferers can become depressed, but they can be medicated for this aspect of their suffering.

The carer feels grief, anger, guilt, shame and embarrassment, and there can be problems with other members of the family who may distance themselves from the situation. Carers need to talk with someone. They should take a break, use support groups, and stop putting the patient before the needs of the rest of the family. One should accept help graciously. Difficulties may arise with a patient who is accustomed to only one carer and only one set of surroundings. I saw this happen with my own father, who became senile in his late seventies, he refused to let my mother out of his sight, and, when we tried to give them both a holiday, he became totally disorientated and had to be taken home. Only she could successfully take him to the toilet.

Dementia cases have only the illness as common ground. The sufferers can come from all walks of life. Some of my patients had led successful lives and they were able to contribute towards the cost of treatment or care, perhaps at Tyrwhitt House. Until I began working in this field, I had little idea of the extent of such cases. I had only ever met my father's case and I assumed that this was highly exceptional. I imagine the Society did not help in many of these cases because it could not provide the intensive nursing required, or families were too ashamed to ask. Not that much could be done for many of the dementia patients.

*   *   *

This was certainly the case with a clergyman I met. **The Reverend Brian Richmond** was disorientated in conversation although he washed, shaved and took himself to the toilet. He showed signs of Parkinsonism, with shaking limbs, and he had had several minor strokes. He was totally unable to concentrate on anything for more than a few minutes. Brian was, however, co-operative and attended a local day centre. Having served in the RAF, he was ordained a priest in the Church of England in his forties. Brian was oblivious to most of what went on around him, but he was able to remember clearly some of the events of his earlier days. Mrs Richmond never completed the forms I left with her, to apply for a place at Tyrwhitt House. I am sure we might have taken him. Doubtless she considered there would be other cases more deserving of help from the Society. That would be her Christian way of looking at things.

I met a retired Colonel from the Queen Alexandra's Royal Army Nursing Corps (QARANC), she was an OBE and her husband was also a retired colonel. Then in her seventies, she had deteriorated rapidly and was totally dependent on her husband. She was pleasant enough but could not read or carry on a meaningful conversation. She had to be dressed, washed and toiletted. The local mental hospital were taking her in for a few days occasionally. After discussion, we all felt that this was the best that could be done in this desperately sad case.

Lionel had been taken prisoner by the Japanese on the borders of India. After the war he trained as a missionary and returned there. He was eventually awarded a War Disability Pension for physical injuries, and he qualified for an Indian Army Service Pension. After a full life of teaching and preaching, it was sad to see him totally incontinent, having to be got up through the night to go to the toilet, as well as being padded up.

Lionel's wife had attended regular meetings of carers at the local psychiatric hospital, where he was taken in for two weeks every other month. I felt that this was the best arrangement that could be made, but I was able to offer her some advice on their financial matters, should he die first.

I could go on to describe many more of these cases where, clearly, the Society could never hope to admit such patients to their homes, generally speaking. Intensive nursing care simply could not be made available for all these sad cases of dementia. The best I could do was to offer to call again, if requested, or to offer advice or support over the phone if matters became urgent. Quite clearly, I was never going to be able to physically devote time to all of them, on a regular basis. At times, I was able to alert other agencies – War Pensions, the Regiment or the RBL – to their needs. A further important reason for not rushing in to admit the clients to Tyrwhitt House was the fear of the patient becoming disorientated. There were one or two worthwhile exceptions, however.

Francis served in the RAF as a bomber pilot right through the war, coming out unscathed. He had a successful career as a stock-broker in the City until he retired at sixty-eight. As can happen, within two years of leaving work, Francis began to suffer from loss of memory and some degree of confusion. He was referred to me by the local branch of SSAFA. His conversation had become a patchy mixture of sense and nonsense. His wife had taken over all their financial affairs, but they were able to go for walks and even play a little golf.

Because Francis' dementia was in its early stages, his wife brought him to Tyrwhitt House for a successful stay. He did not mix with the others, but found the atmosphere pleasant enough, and his wife Marjorie enjoyed a break from her responsibilities. She brought him in and took him home. When he came for a second visit, it soon became clear that he was deteriorating and there was a need for someone to feed him and assist Francis gener-ally for most of the day. Marjorie agreed to pay for an agency nurse, from a medical fund into which they had contributed. This worked on other occasions until the care was no longer adequate. I then suggested that the local hospital might take Francis in for an occasional break, which they did.

Our experience with Francis was to serve as a pointer in one or two other cases, but the cost of such care meant that patients without means could not be assisted in this way. A retired bank manager stared at the wall, but he enjoyed listening to classical music on his headphones. His wife had not enjoyed a break from him for seven years; she looked ill. Their family felt unable to take him, such was their embarrassment. When I suggested that we take their father for a couple of weeks trial at Tyrwhitt House, they agreed to arrange to transport him, collect him promptly if there were any difficulties, and pay for any additional care as might be necessary. We had him at Tyrwhitt House for a fortnight and everything was done for him, padding up at nights, frequent toilets during the day, spoon feeding, and regular medication. His wife was much better for this break, but it had to be said, reluctantly, that this would have to be the first and the last time we could do this for them. The extra nursing care was expensive but the family were able to afford it.

It would have been easy to dismiss the Alzheimer's cases out of hand, but I found that other help, besides nursing care, was needed. Many carers badly needed someone to talk to, and there were cases where friends and neighbours no longer called because they were embarrassed by the unsocial behaviour of the sufferer. My mother was able to keep all her friends, but father, who was confused, did not appear to care what he said or did when there was company present. I could often help the family regarding where to go for some help or advice needed. I found that a friendly face was always appreciated and to call on these cases was never a waste of time. It could be far more pleasant than calling on some of my psychotic clients. There were always a few dementia cases for whom a break in a home was certainly worth looking at, always at the discretion of the Society's medical and nursing officers. However, as time went by, financial restraints became a major obstacle when other agencies became reluctant to assist towards expenses. Nowadays, senile dementia (arterio sclerosis) and pre-senile dementia all tend to be referred to as Alzheimer's disease.

Early in my welfare career, I met one of the pre-senile dementias, specifically diagnosed as Huntington's chorea. **Benjamin** had been a Major in the Regular Army, serving to the age of fifty-five, a soldier all his life. He had last served in Hong Kong where his mental faculties had become impaired, but the Army had permitted him to stay on to complete his full term of Service for his pension. He and his wife bought a house on the South Coast where Benjamin could continue to sail, which had been his main hobby. His memory began to deteriorate further and, when he began to walk with jerky movements and the same movements began to occur involuntarily, he was seen by a consultant.

When Huntington's chorea was diagnosed, this was a severe blow to the entire family. It is known that up to half the patient's children may become affected and, as the disease does not strike until the late thirties, at the earliest, the chances are that any infected children will already have been born when the parent discovers he or she has it. This is how Huntington's chorea is perpetuated because there is no known cure. This illness was first described by George Huntington in 1872 and currently can affect 0.04 per thousand of the population. The patient suffers from a degeneration of the spinal chord and this causes abnormal bodily movements; hence the physical signs occur as well as the general onset of mental deterioration, similar to the other forms of dementia.

The Society took Benjamin into Tyrwhitt House, where he had a successful stay. He was cheerful and took part in some of the activities, however, there were some days when he was worse than others. Preparations were made for a further visit, but on the day I called to collect him he was in hospital, having had a relapse. Bearing in mind that I had only been a welfare officer for a matter of weeks, I was in for a tremendous shock.

I found Benjamin in a high-sided cot where he was restrained by a harness while he tossed and turned ceaselessly. He was emaciated, and Sister said they had encountered the utmost difficulty in feeding him as he was never still. His condition was consuming an enormous amount of energy from him every second. I attended a case conference where it was collectively decided that Benjamin would have to be fed by injecting nourishment into him – an extreme measure. He died a few days later. I felt powerless, as all I could do was to initiate some counselling for the family. They were all concerned for the future of Benjamin's children and grandchildren, then alive and well. I informed the Association to Combat Huntington's Chorea of their predicament.

I was to meet patients with Parkinson's disease, which affects co-ordination of movement and balance; there can be tremor of the hands, slowness of limbs, falling over, and dribbling from the mouth. There is a loss of memory similar to dementia but the bodily movements are not so extreme, as with Huntington's. Parkinson's is treatable using drugs to replace the deficiency of dopamine in the cells of the brain. Thus its progress can be delayed.

One of the side-effects of all these organic states of the brain is depression, and this condition makes everyone else feel depressed, under the strain of it all. The sufferers try to gain attention as a result of their incontinence and immobility, continually demanding help. They feel anxious, tired, gloomy and irritable. They might say they are suicidal, feel inferior and blame others all the time. A patient who has dementia alone seldom complains but, if depression should set in, this can be treated by anti-depressant drugs which usually take about ten days to work. Sedatives merely hide the symptoms.

Epilepsy is not itself a disease but it is often associated with cerebral degeneration. As with the dementias, the sufferer may develop a paranoid personality. Epilepsy is a condition where the nerve cells in the brain have become hypersensitive and fire off uncontrolled discharges, causing a seizure or a fit. It has important legal implications relating to any driving offences committed should the sufferer have a fit at the controls of a vehicle; or if he gets married without revealing to the other party that he or she had epilepsy before the marriage took place. Acts of aggression following or during a fit have led to prison sentences. This condition falls properly within the province of the neurologist, rather than the psychiatrist, and it can be treated with medication. The present state of the law bars an epileptic sufferer from driving until it can be shown that no more fits have been encountered for a number of years. Epilepsy has been treated by neurosurgery. The patients suffer from being socially less acceptable, while their chances of employment are limited through fear of injury or aggression should they have a seizure whilst at work. Thus it can be felt that they are a potential danger to themselves and to those around them.

I encountered the case of **Archie** who had served in the Army, was invalided out following an epileptic fit, and received a 30 per cent pension for a mental condition aggravated by the Service. In fact, had the Authorities known about this condition, Archie would never have been permitted to enlist as a Regular soldier. The local branch of the RBL first drew my attention to his plight. Archie was a physically strong young man, in his twenties, unemployed, living with his parents, and he had been having a fit every other day when I called. On at least one occasion, it had taken three men to restrain him while he was doing serious damage to his parents' home.

When I called again, the seizures were less frequent and he had not had one for some weeks. He was doing odd labouring jobs, and I suspected that his employer was unaware of his history. Archie had no idea of what he did while under an attack. He could not stand crowds and, during my visit, he got out of his chair and left the room several times. We had him registered disabled but all he was officially allowed to do was voluntary work.

We took Archie into Tyrwhitt House, but even there he had one or two episodes of a minor fit, or petit mal. After this, I lost touch with him. Visits to his home found no one there, although I had written to make an appointment. The Society's policy was never to force our presence upon anyone who did not wish to co-operate. There were always many other calls on my time from those whom we were able to help.

The condition of encephalitis arises when a viral infection spreads to the central nervous system. Some recover, but others sustain a history of a lasting intellectual deterioration. Such was the case of an ex-Guardsman whom I found being cared for in an upstairs room at a public house. The Regiment had asked me to call when they had been approached by his carers for a grant; this was to enable them to employ an agency nurse to give them a break one day a week. I found **Kenneth** sitting in his wheelchair and requiring help not unlike any other severe case of dementia, except that he was only in his forties. Kenneth was permanently incontinent and unable to recognise anyone. He was also prone to fits, and clearly beyond the nursing resources of any of the ex-Service convalescent homes.

Fortunately, Kenneth had a full sergeant's pension, after twenty-two years Service, and I was able to provide the regiment with all the information they needed to award the grant requested. The money would be paid following the receipt of regular accounts from the agency supplying the nurse. The sad thing was that, when Kenneth first came to live there, on leaving the Army, he had been fit and well, and able to help in the business of running the pub. The cause of Kenneth's illness was quite unknown, except that it had to do with a virus which had infected him.

Ralph suffered brain damage as a result of military Service, but not active Service. A young man, he lived at home with his parents. He was an NCO in the Royal Military Police, a dog handler in the Far East, when one of the animals scratched him. The wound became infected and Ralph contracted leptospirosis, becoming so seriously ill that the Army flew his father out to be with him in intensive care in hospital.

Ralph recovered, and was discharged with a high percentage War Pension for injuries attributable to military Service. His parents seemed over protective although Ralph's behaviour was near normal. He was friendly, but clearly there was some loss of memory. He was able to use public transport locally but, when we took Ralph into Tyrwhitt House for a successful stay, his father accompanied him on the train. All expenses were paid by War Pensions. When I next saw Ralph, he was becoming anxious over small things, and he worried about his health and his eyesight. It was possible that he was degenerating and losing interest in things. He was, however, able to repeat visits to Tyrwhitt House where he joined in the activities, and gave his parents a break. One wondered what would happen when they were no longer able to provide a home for him. He could become a candidate for admission to Kingswood Grange.

When I first met **Tommy** he was only twenty-six and living on a farm with his parents. He had been awarded a War Pension as a result of a sudden brain haemorrhage whilst serving in the Army, in Germany. Tommy had been an athlete and he had collapsed on the running track whilst competing for his unit. The damage to his brain left him paralysed in his arm and leg, down one side, with some loss of speech. Military hospitals had done their best, but they discharged him from the Army, after a year's treatment.

Tommy was mobile and continent, able to eat, using his good hand. He tried hard to make himself understood. There appeared to be little to motivate a young man in this condition, but we were able to take him to Tyrwhitt House in order to give his parents a break. There, he was able to learn to play snooker and his independence generally improved. Understandably, Tommy later became bored with his life and I felt that more might have been done to motivate him at home where he took no interest in the farm. The War Pensions Handicrafts Officer was considering what might be done for Tommy. I would have thought that something in a sheltered workshop might have been possible as Tommy certainly understood what went on around him.

* * *

My final group of patients in this chapter are those whose illnesses were thought to be psychosomatic, associated with psychological factors, or simply just in the mind. Any part of the body can be affected by mental illness and a situation can arise where the GP tries to treat physical symptoms, perhaps unaware of a psychological root cause of the illness. Many conditions can arise as a result of depression, mental stress or nervous anxiety. They might include digestive disorders, even a peptic ulcer, asthma, hypertension, eczema, urinary and sexual problems, and migraine. This list is by no means complete. It has not been firmly established why any particular area of the body might become affected in this way. These physical conditions actually exist and are not figments of the patient's imagination. These contrast with a second type of situation where there really is nothing organically wrong with the patient, but the pains he encounters certainly exist. However, their cause may be real or imaginary. Much time was devoted to some of these clients of mine, not always with any success beyond my attempting to provide some degree of reassurance. I should mention that a patient with a mental illness can be a very vulnerable person and even minor ailments can be magnified out of all proportion. This can be an aspect of attention seeking, or finding something to complain about. It can occur especially when the client is lonely and welcomes a visit for any reason at all.

* * *

Douglas was called up after the Second World War and was discharged three years later with duodenal ulcers for which he received a War Pension. Depression set in, but he was able to do labouring jobs. His wife left him because of his moods and, when I first saw Douglas, he was bored and tired of the colourless diet he had to endure. In addition, he had developed a rash all over his arms and legs. He lived alone. Although he received some treatment from a dermatologist, and we were able to give him medicated baths at Tyrwhitt House, his wounds continued to weep. The ointment he was prescribed ruined Douglas' clothing which we replaced where possible. His condition came and went in terms of severity, but it never cleared up completely.

Joseph served right through the Italian campaign, coming out with a War Pension for psychoneurosis, caused by battle fatigue. Every time I called, he worried about whether he was receiving the correct amount of money for his pension and allowances. Joseph developed a heart condition which affected his limbs. A visit to the dentist, and the anaesthetic used, was said to have affected his heart further and, although he recovered after a period in intensive care, he was left with pains in his mouth which never went away. A major hospital, after thorough investigation, stated that these pains were psychosomatic. They suggested hypnosis but it never worked. We had Joseph in Tyrwhitt House to give his long suffering wife a break, but he died soon after from a heart attack.

A local GP once asked me to see **Paddy**, whom I found walking everywhere with a stick, even in the house. He complained of back pains, and was continually standing up and sitting down, holding the base of his spine. Paddy's back trouble was attributed to a fall two years earlier and he had not worked since. He said he could not go to work on his bicycle because he kept falling off!

He denied any mental illness, and blamed all his troubles on the side-effects of the drugs he had to take for the pains in his back. They gave him blurred vision and dizziness. Paddy had done his National Service in Singapore and came out physically fit. We had him in Tyrwhitt House and, afterwards our Consultant corresponded with his GP. Paddy was given exercises to do but the pains in his back persisted. We eventually had him X-rayed at a major London hospital on one of his visits to Tyrwhitt House. Nothing was found to be wrong with his back, and the staff at Tyrwhitt House told me that there were occasions when he got up out of his chair without using his stick. Paddy returned there many times. He always complained of his back, and never went back to work.

Maurice suffered from depression. He had served right

through the Second World War and had a high percentage War Pension for psychoneurosis. He told me he was allergic to anti-depressants, so he had to suffer and make the best of things. I found Maurice sceptical in outlook, and his wife said he could be difficult to handle. When he complained of pains in his throat and neck, his GP said it was all psychosomatic and that nothing could be done for him.

At this point, Maurice's wife took him down to the surgery by taxi, dismissed the cab, and refused to take him home until he had been given a thorough examination. The doctor diagnosed poliomyelitis and had him rushed to hospital. There it was found that he did not have polio, but cancer of the throat. After treat-ment, Maurice recovered and felt better than he had done for many years. The hospital took over his condition; his headaches were treated and cured. So much for Maurice's pains in his throat being psychosomatic.

The outcome of this case was that the Society appointed a local psychiatric consultant to whom we could in future refer patients, such as Maurice, where a GP was clearly unable to recognise our client's problems. As a result, we were to obtain better co-operation from GPs in that area, and I felt that Maurice had not suffered in vain.

In the country generally, GPs appreciated what we tried to do to help their patients, and the Society's Medical Officer, himself an eminent GP, was of tremendous help in promoting good relationships.

The penultimate chapter will deal with a summary of the whole range of problems affecting the mentally disabled ex-Service person today, and how some of them are being resolved or left unresolved.

Note

[1] PTSD has been thought of as a negative concept. Post traumatic stress reactions can include not only PTSD itself but also post traumatic stress depression and post traumatic anxiety states other than PTSD, such as panic disorder and phobic disorder. In fact it can be a life-saving device, an adaptive mental process, a method of coming to terms with a threat to life.

11

What Can and Cannot be Done for the Mentally Disabled Ex-Serviceman

I have attempted to show that there is more to being a Welfare Officer than simply sending clients to Tyrwhitt House and concentrating on the War Pensioners. There is a whole range of sick people seeking support, as depicted in the last chapter, but appropriate help will depend largely on the age group concerned. For the younger person, employment is perhaps the most significant gift that could be offered. Yet, because our clients are mentally ill, some are unreliable and employers are naturally apprehensive. We no longer enjoy full employment for the able-bodied, but even if we did the average firm could not afford to take on staff who did not turn up regularly.

The self-employed have been my most successful clients; the next most successful have been those fortunate enough to work for a boss who was himself ex-Service, or in a sheltered organisation such as Remploy. Not enough work is available generally for the handicapped. The Royal British Legion has a scheme for assisting ex-Servicemen to set up in business, but at least one of my clients ran into difficulties caused by his mental state, and he had loans which could not be repaid. The idea of Milner House, with its own workshops making a marketable product, was very sound in its day, between the wars. But it could never succeed in today's competitive world of business. The RBL has a scheme for goods made at home by the disabled to be sold in their clubs, but much more would have to be done to make such a scheme provide worthwhile occupation for some of the clients I have met.

Apart from clients who worked for an ex-Service employer, others who did well had an understanding wife who was able to stand in when the man was ill. Working long hours indefinitely is not always possible for a manic depressive or an epileptic. One is tempted to look back to a 'golden age' of the large mental hospital with its own farms and gardens where patients were both supervised and motivated. Some had a shop selling craft work to staff and visitors. The patients I saw ten years ago all received wages for their work and wore their own clothing. Wearing hospital clothes and working for nothing had long since disappeared. Therefore the large hospitals were not so bleak as they were sometimes depicted. It should be remembered, however, that it was only in the 1950s that these institutions opened their doors allowing patients to go outside, having benefitted from the recent advances in drug therapy. The horror stories of patients locked inside, in brutal conditions, had often been true and some of the arguments to close the asylums stemmed from a misguided belief that brutal regimes were still the norm. Nevertheless, there were patients who should not have been there in the first place.

The closure of many of the military hospitals, with psychiatric departments, in April 1996 added to the problems of my clients. During my time, invaluable help was given to the clients referred to the RAF hospital at Wroughton near Swindon. Then there was the Cambridge Military Hospital at Aldershot (named after HRH the Duke of Cambridge, cousin of Queen Victoria and Commander-in-Chief until he retired in 1885, aged 76!); also the Queen Elizabeth Hospital Woolwich which had replaced the Royal Herbert Hospital in the 1960s (named after Sidney Herbert a contemporary of Florence Nightingale). The military hospital at Millbank had been closed by the 1980s as had Mill Hill Hospital in North London. There were many others which had been closed since the Second World War, including the Military Psychiatric Hospital at Netley, Southampton.

War pensioners had enjoyed some rights as priority patients in military hospitals, for their accepted disabilities, and these hospitals were regarded as part of the family of ex-Service facilities. They were friendly institutions but they were much more than that for the mentally ill. We knew that the psychiatric medical and nursing staff were sympathetic to the Society and appreciated the circumstances in which our patients had served their country. I feel these losses can never be replaced and the expertise they contained has been dispersed. It is true that some fragments of the military medical Services remain, including a military wing at Camberley and the Royal Naval Hospital at Haslar. These survive intact, concentrating on the needs of serving military personnel in the South of England. Where hospitals have closed, the military medical staff who continue to work as doctors and nurses will probably join the NHS or private medicine, where the vast majority of their colleagues will have very little idea of what Service life is like. All this is in addition to the increasing age difference between our older clients and the medical staff of today. Handling mental cases calls for creating a basic confidence in the patients, and this will be all the more difficult to achieve in the future. Many of my cases were those where the patients responded to a military hospital regime and expertise far better than the local NHS facilities. Many had tried both. This is no disrespect to the latter, but these are facts of life, as I found them.

These problems are by no means insoluble given the availability of the most suitable staff for our clients. The nurses in the Society's homes are mainly no longer ex-Service. The War Pensioners' Welfare Service has operated very successfully in recent years, recruiting staff with no previous Forces experience. I feel that this success has been achieved largely due to the skill with which their officers have been selected. Perhaps this is a lesson to be learned by all who look after the ex-Service community.

Thus we have a considerable pool of retired Forces medical and nursing staff, with psychiatric training and experience, who ought to be known to those who will continue to look after the mentally disabled ex-Service community. This is particularly relevant to the Tribute and Promise we made in the summer of 1995, marking the fiftieth anniversary of the end of the Second World War. There were public ceremonies all over the country at which the following Act of Commitment was made:

"We, the voluntary organisations of Tribute and Promise
pledge ourselves anew to our work in support of the
wartime generation.
We promise to do everything possible to help where
there is a need, and to ensure that they may enjoy the years
which lie ahead in comfort, dignity and contentment.

> This we promise as a lasting token of our appreciation
> and gratitude."

The long list of participants included:

Air Crew Association
Army Benevolent Fund
British Commonwealth ex-
  Services League
British Red Cross Society
Burma Star Association
Dunkirk Veterans' Association
Fire Services Benevolent Fund
Gurkha Welfare Trust
Italy Star Association
King George V Fund for
  Sailors
Merchant Navy Welfare Board
Far East Prisoners or War
  Association
NAAFI
Normandy Veterans
Not Forgotten Association
RAF Benevolent Fund
Royal Naval Association
SSAFA
St Dunstan's
Royal British Legion

War Widows' Association
WRVS Trust
Age Concern
BLESMA
British Korean Veterans'
  Assoc.
Carers' National Association
Ex-Services Mental Welfare
  Society
Corps of Commissinaires
Forces Help Society and Lord
  Roberts Workshops
Haig Homes
Help the Aged
The Order of St John
Rotary International
RNLI
Salvation Army
Scouts' Association
Townswomen's Guild
YMCA
And there were many others.

Milner House once had a market garden as well as the Thermega Factory. Tyrwhitt House used to cultivate its own land, producing vegetables for the home. But this was found to be impracticable as patients came and went. There were too few staying with any interest in such work. In my time, some clients were tried out on doing jobs repairing and redecorating the house, but, being a short-stay home, some patients left leaving work unfinished. The Occupational Therapy Department always had potential to make articles for sale, but, again, production could not be sustained for long when patients changed and new ones arrived. Many of my clients were over retirement age but I feel more might still be done to occupy, or even employ, the younger men and women, in spite of their mental problems. I have had clients who worked in Remploy factories around the country, but in recent years this splendid organisation has had difficulty in finding sufficient work for employees who are mentally fit although physically disabled.

We are therefore victims of the continued progress of the computer age and of the large unit of production with economies of scale; and there are cheap foreign imports of the kinds of goods the less able might have made in this country. The taxation and benefits system currently operated works against clients on a low wage. One of my patients, a minibus driver, struggled to remain at work rather than doing nothing. When he had paid all his dues, he had very little left. You see, his friends at work all did overtime to make up a living wage. My client could not manage this because, after a day on the road, he was mentally exhausted because of his condition. Unfortunately, we could not get him a War Pension. Even our War Pensioners, in some areas, as has been said, have their benefits cut by the Local Authority as a result of the pensions they receive.

A few more case histories will illustrate some of the problems arising from Care in the Community, or rather the lack of it. In spite of all the facilities available, some clients are slipping through the net, for lack of resources.

*   *   *

Stanley joined his Regiment at seventeen and a half and served six years before taking his discharge. He probably knew he was not well. He tried various jobs but became increasingly depressed following the loss of both his parents. When his wife left him, he made two suicide attempts. At one stage he was totally unemployed for three years. Stanley now lives in a council flat and receives help from SSAFA to pay off an accumulation of debts. He relies on anti-depressants and it was SSAFA who asked me to pay him a visit.

I tried for two years to get him into Tyrwhitt House for reassurance and a psychiatric assessment, but the Local Authority would not contribute. He is mobile, continent, and capable of work. He is well known to the local social workers in Mental Health, and they give him voluntary work to do, but nothing ever develops from this. Stanley is barely coping on his own in his poorly furnished flat, and he needs clothing. He told me he can't afford to have his heating on much. There is a history of him stopping his medication whenever he feels a little better. This is usually followed by urgent admissions to hospital. Meanwhile the social workers come and go.

Stanley would have been a candidate for Milner House in the old days, or in one of the old mental hospitals as a temporary patient doing sheltered work. He does not expect much from life, but I would say he is getting very little from Care in the Community. The Local Authority never did pay towards a stay at Tyrwhitt House. Very recently, we were able to arrange for him to have a week's holiday at Bournemouth, at Cedarwood Guest House, with his two young sons, paid for by SSAFA and the Not Forgotten Association. He left before the week was up because he was unsettled.

At the other end of the scale, socially, **Brendan** was an outstanding pupil at school before joining the family's wholesale business. He was called up in 1942 and was commissioned in the Middle East where he rose to the rank of Major. He was a Staff Officer in Lebanon, Palestine and Cairo. After the war he went back to the family business, which he later bought outright. He retired at sixty-seven. When I met Brendan, he was in his eighties, suffering from loss of memory after a fall. He was unsteady on his feet and spent two days a week at a day centre. He was feeding and dressing himself, very deaf but cheerful. When asked to contribute towards a stay at Tyrwhitt House, because the local Social Services would not help, he refused to pay, claiming that he had paid towards the Welfare State all his life. His wife badly needed a break but here was a stalemate position. Brendan could probably have afforded to go to a private nursing home, but he would only consider Tyrwhitt House because he liked the idea of being with ex-Service company. Reluctantly, we decided we could not help.

Terence lived alone in a bare flat. He wrote to the Society, in a letter hardly legible because he was blind in one eye and only partially sighted in the other. His wife had died twenty-five years previously. The flat had only one chair and a bed; there were no carpets or curtains. Terence had some memory loss but he was mobile. He managed on just the basic OAP. He had served in the Royal Navy, and took part in the D-Day landings, but he was later discharged through loss of his eyesight, which he said was not attributable to the War. For the next twenty years, he was in and out of mental hospitals; it might have been better if he had been kept inside and quietly occupied. Terence was taking anti-depressants and sleeping tablets and he was seeing a psychiatrist.

When I made contact with her, Terence's social worker said she had never seen the inside of his flat. She only called the once, and that was to take him to look at an old people's home. When he declined to go, they washed their hands of him. They refused to pay for Terence to come to Tyrwhitt House so, once again, we had a position of stalemate.

One wonders just how many more cases there are like Terence. Perhaps now that there are more welfare officers employed by the Society than there were in my day, it will be possible to follow up some of the cases I had to abandon.

I have already remarked that we do not know the extent of the problems of mentally disabled ex-Servicemen and women, in terms of their numbers or the range of their disabilities. One thing is certain, numbers are not going to decrease by much in the foreseeable future. As an indicator only, the report published by the War Pensions Agency, for 1995/6 revealed the figures detailed below:

Disablement Pensions in Payment	1995	1996
Total pensions in payment (disablement, widows and others)	309,841	323,745
Disablement pensions alone	260,297	265,375
War Pensioners' Supplementary Allowances	206,561	204,471
Claims received	99,081	70,369

As has been long predicted, overall numbers of War Pensioners are now beginning to decline slightly. The comparable total figure, for 1998, now stands at 317,659.

War Pensions open days, where the public have been invited to come along and claim for Service injuries, have undoubtedly led to an increase in cases over the past few years, together with the efforts of the Society, the RBL and other charities working with the disabled. The backlog of undiscovered cases may now be expected to decline, but the fact remains that we still have over 300,000 war pensioners whose interests will need to be served for the foreseeable future.

My experience was that War Pensioners constituted about half my case load. Thus there is a similar number of mentally sick clients whose illnesses were not recognised as having been caused directly through Service in the Forces, and yet who may need help badly. Sometimes, their need is more urgent than those receiving War Pensions as they are only getting the basic allowances from the Welfare State, and sometimes not even this. Today's problem is that they now depend on the goodwill and generosity of the local authorities, and the NHS, and clearly, there are not sufficient resources to cover their needs. The voluntary bodies are stretched to the limit, and the number of appeals for help from these clients shows no signs of decreasing.

No decision has been reached by Government in response to a private member's bill from Simon Hughes, Liberal Democrat MP for Bermondsey, which would make it obligatory for all local authorities to disregard War Pensions when calculating an assessment for Council Tax and Housing Benefit. This is most desirable but it would only benefit those already in receipt of such pensions. It would not affect those thousands of others with no War Pension at all, and these are often the most needy cases.

It is now possible to sue the Ministry of Defence for negligence, just like any other employer, where they fail to provide safe working conditions, vehicles, Service equipment and the like; or if they provide inappropriate medical treatment, or there is proven negligence on the part of the Service Medical Authorities. The RBL suggests that all claimants should approach them for assistance in making a claim. This idea of compensation was successfully taken up in the case of a Falklands veteran in the Scots Guards, mentioned previously.

The fact remains that the Parliamentary Commissioner Act of 1967, setting up the present system of appeals to an impartial ombudsman, does not apply to Servicemen. Servicemen have always surrendered many of their rights on enlisting, in the interests of military discipline and national security. The suggestion has been made that there should be a Department of ex-Service Affairs, under its own Minister, a member of the Government, who would handle all cases of complaints against the MoD. I came across cases where a Serviceman, through no fault of his own, or in a situation of considerable doubt, was penalised on discharge by an unfair employer's reference. There is no redress for this and consequently, such persons are unable to obtain employment without considerable difficulty. There might even be temptations to embark on crime, where a previous work reference would be of no account! Such a situation could only make a client's mental condition worse, or might even be the cause of it. The RBL continues to make representations on this matter, pointing out that there are departments of Veteran's Affairs in Australia and other countries. The official reply, thus far, has been that the creation of an additional Department of State, with all the ensuing bureaucracy, would not help matters. There are many who would disagree. Again, this would not help those veterans whose disabilities were not directly caused by the Service authorities.

The RBL has attempted to uncover some of the problems we face with the ex-Service population today. The National Chairman reported to his members, in October 1994, that one-quarter of all the country's single homeless are ex-Service. This does not include National Servicemen, but it does include Merchant Navy seamen. Eighty per cent are over forty years of age. The report concludes that the Legion could not possibly resolve this problem alone. Housing is generally outside its scope, but the Chairman suggested creating some half-way house type of accommodation to give some of the ex-Service homeless a base. Handing over redundant married quarters would not be a solution, as many are in areas where there is no work, or they are in need of substantial repairs.

The Legion also has a scheme for assisting towards the costs of residential care in a home for the elderly, or to obtain care for those who are sick or disabled in their own homes. It was revealed that it now costs on average £14,000 a year for a room in a residential home, while Local Authority home helps cost about £8 an hour.

Hospital care remains free under the NHS but all other help has to be paid for, and is means tested by the Local Authority. The Legion offers an insurance scheme to provide £500 a month towards residential home fees for up to three years, or up to £500 a month for help at home for up to three years. Premiums increase depending on the age at which contributions start. This is not a complete answer but it could help many who otherwise would have to sell their property immediately in order to pay for care. The very existence of such a scheme should alert the public to the very real problems of those who approach old age.

The Royal British Legion introduced full-time County Field Officers in 1994 in order to oversee the work of volunteers, on a county basis, covering both welfare work and fund raising. This measure, together with the new Training College at Tidworth, are but two substantial ways in which the Legion is moving forward, away from its image as an association of inwardly orientated social clubs.

Care in the Community took a new twist in April 1993 when funding for the elderly and the disabled was transferred from central to local government, as has been mentioned. The aim was to encourage as many people as possible to stay in their own homes. Very soon after the new rules took effect, the Forces Help Society Princess Christian Homes discovered that local authorities were reluctant to sponsor anyone to stay there if it could possibly be avoided; even those already sponsored for residential care were placed in council-run homes. The authorities had a duty to their ratepayers to maintain their homes in full occupation, if only for reasons of economic management.

My own experience has been that ex-Service people are always happiest in a familiar ex-Service residential environment, where they have much in common with the other residents. This is one aspect of caring for the disabled and elderly ex-Forces population that ought never to be lost sight of. After all, we all signed the Tribute and Promise. The happy atmosphere in the ex-Service homes I visited was there for all to see; this is more than can be said for some of the private or Council-run establishments I came across.

There have been other developments as a consequence of the changed situation in April 1993. In order merely to operate the new system, the Society has found it necessary to replace the part-time clerical assistants in all the regional offices by full-time office managers. This has been in order to comply with the newly-imposed local government bureaucracy, thus adding considerably to the Society's running costs. The smaller ex-Service homes run by some of the other charities now face a crisis whenever they are unable to fill all their beds. Meanwhile, in order to remain in them, some of the elderly residents have been obliged to sell their own homes; and there is no guarantee of them remaining where they are when all their money runs out. These problems were highlighted by SSAFA in 1994, and as a result many charities and their patients are in a state of uncertainty about the future. This situation is surely wrong.

Many who stick it out in their own homes are lonely and have very little to look forward to. These problems are compounded when the clients are mentally disabled; it is right and proper that they be cared for in an ex-Service home. Not that all the ex-Service homes can cater for those with mental illnesses, that is, with the exception of those belonging to the ex-Services Mental Welfare Society.

SSAFA in 1992 reported a 4 per cent increase in the overall number of cases they handled. Welfare visits increased by some 11 per cent, while financial grants to those in need went up by 25 per cent from SSAFA funds alone. Increased discharges from the Forces had led to an increase of 21 per cent in housing and reset-tlement work. This may have been a temporary situation, but ought not to be ignored. It was not surprising that SSAFA in 1992 ran at an overall deficit, and the cost per case and the numbers of volunteers employed all showed considerable increases. Clearly, needs are still there; and here they were, on the whole, dealing with younger ages of ex-Servicemen as a consequence of the run down of the Forces after the Cold War, and the failures of the welfare state to meet all requests for temporary help.

Psychiatric medicine continues to make advances, and these are of direct relevance to mentally-ill clients. It is relatively recently that many in care have moved away from the large mental hospitals, and indeed most of the Dickensian features of them had disappeared long before the reformers began to empty them. These idealists have even fed us with the idea that they should be abolished altogether. Perish the thought! The community today is not a caring place. Nor is it cheap to run a large number of smaller establishments. The local community often objects to such homes being set up, unless they can be built in someone else's back-yard. Such rejection can follow the mentally ill or the retarded frequently; few neighbours are prepared to befriend these patients, for fear of affecting the tranquility or the tone of the locality. There are few sheltered workshops for them, and day centres for the mentally ill are often lacking in facilities. In an NHS hospital the patients get three good meals a day, and there is some social life, as there was in the old mental hospitals. In the case of general hospitals today, the psychiatric wing might be regarded as a 'Cinderella' unit, and there is still pressure around for some of them to be closed.

Today, many who genuinely need asylum are denied it. In the past there was always work available within the therapeutic community. Many of those discharged into the world outside do not take their medication – forgetfulness can be part and parcel of their illness. I have seen these situations for myself. On the other hand, they can overdose where they lack supervision, or become suicidal. Society needs more community psychiatric nurses; they are the key to any successful operation of the Care in the Community policy. The quality of those nurses I have met has been outstandingly good. The failures of the system can end up in Salvation Army hostels for the homeless, or in some places which can only be described as doss houses. Some are in prison; some are sleeping rough. None of those unfortunates I saw had any money. Many of these clients, because of their illnesses, had no notion of how to manage their affairs. Only a few, I found, were integrated within the local community, and were happily doing some work. Some I knew just sat in the park for hours on end, unable to join in local activities such as clubs or day centres, if indeed there were any.

There is a need for rehabilitation therapy for psychiatric cases and this is not often found in general hospitals. These patients do not need beds in the daytime, instead they require space for work experience and constructive recreation. Hospitals in city centres are simply not suitable for mental patients who require more than a very brief stay. Psychiatric day hospitals are not often adequately endowed with facilities; there should be occupational opportunities together with access to the CPN and a psychiatric consultant. Social clubs for the disabled and sheltered workshops all need to expand. We have reached the stage where the closure of some of the mental hospitals ought to be reversed, if only because there is always going to be a hard core of patients who simply ought not to be released, and Broadmoor is not the answer. I know this because I have tried to intervene on behalf of some of these clients. The small residential unit is not large enough to provide the environment and the therapy the large hospitals used to have.

Medical research continues, and advances in biochemistry could produce a breakthrough in our understanding of schizo-phrenia and the mechanisms of the brain itself. Schizophrenia can reoccur in patients who have had treatment and it can be found in those found sleeping rough in the inner cities. Some longer-stay units are still going to be needed for them. A more specific anti-dote for the affective psychoses, such as manic depression is still being sought. And there is on-going research into the circum-stances which trigger off episodes in the patient who is depression prone.

The psychoneuroses are being treated by GPs, social workers, psychotherapists and many others, but all they tend to do is to bring temporary relief. Some improvement in the home circum-stances may be a preventative measure, but ought we to interfere with the patient's right to privacy? New drugs arrive but there can be side-effects such as addiction. Anti-social behaviour can follow a temporary release from inhibition. On the other hand, an increased use of group therapy can enable more sufferers to be reached.

Dementia continues to be investigated with a view to estab-lishing its root cause. This could lead to preventative treatment, and this would be a major advance in medical research.

Hugh McManners' book *The Scars of War* describes the con-dition of PTSD as an inevitable result of active Service for many soldiers, both serving and ex-Service, today. Unless treated, fear of fear will lead to uncontrolled behaviour. If the sufferer can gain insight he can become stronger. 'Pull yourself together' is no help at all. Guilty feelings, frights and nightmares are all signs, yet

Service colleagues, who should know better, think that going on a stress management course at Haslar is going on a 'skive'. There are some who do not attend for fear of missing out on promotion. The 'Stiff upper lip' tradition persists and can deter men from seeking treatment for PTSD. Unfortunately, it does not pass away with time; it is interesting to note that some Jewish doctors in the UK were amongst the first to recognise this condition, having first seen it in the victims of Auschwitz. It is only in recent years that the Service Authorities have recognised PTSD as an illness. The Service Chiefs have been more preoccupied with peace-time problems such as fighting the cuts in Defence expenditure. It is still alleged that some of the sufferers were substandard soldiers anyway. Many who cracked up were, in fact, vainly striving to maintain the 'Stiff upper lip'.

There was some talk of a PTSD Centre being set up, but its opponents argued that this would only encourage the weaklings to opt out of dangerous postings by attending for treatment. The Falklands presented a golden opportunity for psychological debriefing to be explored, but it was missed. Perhaps there was a fear of paying out large sums by way of compensation; and, what about its detrimental effect on recruiting? Thus we have gone right back to the arguments prevalent at the time of the First World War. The Services have been reluctant to call upon psychiatric support because it has been felt that their personnel selection, training methods and deployment tactics can be shown to be imperfect. The Ministry of Defence perhaps feels threatened and yet the military psychiatrists should be instilling confidence towards saving highly trained, expensive personnel for future use.

Thus there is no shortage of clients seeking help from the ex-Services Mental Welfare Society, the only organisation of its kind – perhaps anywhere in the world. If this book succeeds in highlighting some of the Society's achievements, it will not have been written in vain. I shall always be grateful to the Society for enabling me to work for the mentally disabled ex-Service community.

When I joined the Society in 1984, there were two Welfare Officers based at Head Office in Wimbledon, and together we covered the whole of southern England and Wales. An office in

Manchester had opened in 1953 where one officer covered the whole of northern England and Wales. There were two officers in Scotland and, from 1982, another in Belfast for the whole of Ireland. Because there were so few of us, we travelled thousands of miles a year, often only visiting the most urgent of cases. At most, my clients could only be seen once a year. I sometimes drove a thousand miles in a week.

As the number of cases increased, it became necessary to increase the number of Welfare Officers; each new appointment made led to an even greater increase in the number of clients. My colleague at Wimbledon was enabled to set up his own regional office, for East Anglia, in 1987, and his desk was quickly occupied by a new officer to cover London inside the M25, but he too was soon overloaded with new cases. The Manchester Office acquired a second officer, as did Glasgow, while the South West had its own regional office under a new Welfare Officer who arrived in 1988.

By this time, Tyrwhitt House was clearly not large enough, even though a second Society home at Hollybush, in Scotland, had opened in 1985. Tyrwhitt House was extended in 1993, having been previously enlarged in 1979. A new regional office for the West Midlands opened in Worcester in 1994. Thus the Society became overloaded by the success of the work it did. I recruited new clients every week, far more than the numbers who died; I saw 500 of them every year and travelled 30,000 miles to visit them. The tremendous achievements of the Society produced its own problems, even to continue to provide the Service it did, without extending its scope in any way. All of this again illustrates the point that we do not know just how many potential cases there are. Care in the Community and Local Authority funding, or lack of it, created an increased administrative load requiring improved support for the welfare officers at the same time as new branch offices were being opened. More recently, a second welfare officer has been appointed for Central London. When I covered most of this area, I knew I was doing little more than merely scratching the surface of the problems in a haphazard manner; I could do very little else, given the time I could devote to the Capital.

Respite care at Tyrwhitt House has been unique to the Society, in that specialisation into the care of mentally ill ex-Forces personnel was not available anywhere else, outside the military hospitals. As has been stated, treatment in military hospitals, appropriate though it was, almost disappeared completely in 1996. The Society's homes are full, yet there are substantial numbers of clients, who deserve treatment, who cannot be

Audley Court, Newfort, Shropshire, was opened in July 1997.

The reading room
at Audley Court.

The games room
at Audley Court.

admitted because no one will pay for them. Before 1993 Central Government paid the Society direct for all cases, and the administration required was minimal.

The Society has had to concentrate its efforts towards those in receipt of War Pensions for psychiatric conditions and who, consequently, can be funded by the War Pensions Agency. A few others have been admitted who could afford to contribute themselves, or who could find a friendly Local Authority or some charity to help. If this book is successful, I would very much like to devote the proceeds towards financing poor clients, not otherwise assisted, to come along and receive treatment in the Society's homes. The most recent development of all has been the establishment of a new short-stay home in Shropshire; this should take some of the pressure away from Tyrwhitt House and Hollybush House. No doubt it will soon be running to capacity – all the signs are there. The War Pensions Agency is empowered to fund clients in the Society's homes for up to six weeks a year. Few are ever admitted for this total length of time because of the number of clients competing for places at any one time. The homes continue to offer, besides excellent food and accommodation, occupational therapy, group therapy, counselling and the Services of a psychiatric consultant. Additionally, there is basic nursing, some entertainment and social activities including excursions by minibus, and games. All the homes are set in tranquil surroundings with opportunities for walking and other forms of exercise.

A second extension to the Society's permanent home for veterans at Kingswood Grange was opened in 1990. But, again, more facilities of this kind could be filled, if they were available. This is because other ex-Service homes cannot usually take the kind of clients the Society was set up to help. The alternative can only be private homes for the mentally disabled where the ex-Serviceman does not always feel comfortable. Ex-Servicemen, especially when they are ill, thrive in an atmosphere where there are other patients from similar backgrounds. The comradeship of those who have served in the Forces, even for a short period, is a unique concept which may have seen them through difficult situations on active Service; that comradeship can still be there even in later life, and can be a great comfort to these people in all kinds of adversity later on. The Society has confirmed this fact of life over and over again by its success in all its homes.

Milner House was sold in 1990, as it had outlived its original purpose of providing a home and work for young ex-Servicemen. Towards the end of its 'life', nearly all the occupants were over state pensionable age, while the factory, then under Remploy, was employing physically handicapped workers, with only a handful of the Society's clients still on the payroll. It has to be accepted that finding work for the mentally disabled today is not easy, when competition for jobs among the able-bodied is so keen. Even the RBL, with decades of experience in employing the disabled, cannot provide work for all who apply to them today.

All the occupants of Kingswood Grange are elderly, so the possible objective of finding work for them does not arise. A home for the younger mentally disabled ex-Servicemen would probably not be feasible today, if only because numbers are relatively small as the size of the Armed Forces themselves continues to decrease. Were such an idea to become a reality, the Society would undoubtedly be the best organisation in the world to run it. One

Summer Fête at Kingswood Grange, 1997.

is tempted to attribute the success of the Society today, partly to the fact that the Community seldom cares for the mentally disabled. In the days of the large hospitals, they could be out of sight, out of mind. Relatives might abandon them, but carers often are in need of respite themselves.

To employ clients in the Society's homes, which was once done, would only be a drop in the ocean of unemployment. Repeatedly we come back to the fact that the mentally ill can never guarantee to work continuously – they have black days. Awarding pensions may be a means of recognising a man's bravery, it helps him feel secure financially, but it does not fill his days with useful occupation. Among my happiest clients were those with absorbing hobbies, be it gardening for themselves or for others, or making things for sale or for their friends. A client who tried to be a commercial representative could not keep up with the demands of the work to make a satisfactory living out of it. The most unhappy were those who just sat in front of the television all day; some of them did not live long. Clubs and day centres for the handicapped have proved useful, but many are full of civilians with little or nothing in common with my patients. Even the social clubs and centres designed for ex-Servicemen are not always appropriate for a client who may suffer from episodes of unsociable behaviour. Only the Society's homes are generally able to cope with these situations.

It has been suggested over the years, that the Society might open its doors to members of the Emergency Services found to be suffering from illnesses caused by stress. Certainly, the expertise is there and there have been situations where the Home Office has supported treatment for Police and Fire Service personnel at RNH Haslar. The Society has treated such cases, but only where the patient has also been an ex-member of the Armed Services. Two reasons for not following this line of action would appear to be that to accept these civilians, although very deserving, could open the gates to a flood of applicants with which the Society could not cope. Secondly, the end result might be to change the ex-Services character of the organisation which has been a unique source of strength and a significant reason for its success.

Thus, the ex-Services Mental Welfare Society has a distinguished record in dealing with the mentally disabled ex-Service population of un-numbered thousands. There is no end to the volume or scope of the work which remains to be done for them; whether we could or should develop further into the employment scene remains an unresolved issue. Perhaps there is no satisfactory solution to be arrived at today because suitable work for the mentally disabled is just not available, except as an aspect of occupational therapy. It is significant that the most recent improvements at Tyrwhitt House, in 1996, have included a complete new occupational therapy building.

Looking after the most seriously disorientated ex-Servicemen, the dementia cases, has been considered inappropriate for the Society, if only because, in most of these cases, the patient scarcely realises where he is. Therefore there would be no advantage in placing him in an ex-Service environment any more than any other place. The Society has always drawn the line at patients requiring intensive nursing care where costs alone would be prohibitive for a charitable organisation, without massive support – which is not likely to be forthcoming from Public funds.

The Society could provide more for the needs of the disabled non-War Pensioners, who are often the most deserving of all our clients, and most in need of the convalescence. More non-pensioners could be helped if more funds were available. If the proceeds of this book can help here, even in a small way, I will have made a positive contribution towards resolving a problem I have seen for myself. I could name many clients of my own who could benefit from such a scheme, if one were set up.

There is a pool of unique and valuable expertise in the care and treatment of the disabled ex-Service population and which remains largely untapped today, when it is most needed. I refer to the dozens of doctors, nurses and others formerly employed in the psychiatric departments of the military hospitals closed in 1996. Some will have taken retirement, others will be in private practice or in the NHS. I am also aware that some retired staff, formerly employed by the Society, could make a contribution in a field where appropriate training and experience are not easily obtained. The Society has a small number of psychiatric consultants employed today on an ad hoc basis, and most valuable they are. Such a scheme might be expanded to include not only consultants but also psychiatric nurses and welfare officers, both paid and unpaid, depending on how they are used. So long as the size of the problem of visiting, assessing and treating mentally disabled ex-Servicemen remains unknown, these untapped reservoirs of excellence ought not to be forgotten.

12

The Lighter Side of the Work

Without in any way trivialising or diminishing the work of Welfare Officers, the most valuable asset one can bring to the job is a sense of humour. And this held true whether one was working alongside colleagues at Head Office, where I was based for seven years, or knocking on doors of potential clients.

One client, who often phoned Head Office in Wimbledon, claimed that he had been knighted by Sir Winston Churchill for Services to the country so secret that he was not allowed to tell us what they were ! 'Sir Fred' also promoted himself to imaginary senior ranks within the Service. We all saw the funny side of this. Tact was of the essence when ministering to Fred's needs, but he certainly appreciated being able to telephone in with his problems. When I was out, the ladies in the office were most kind to him. Fred tended to wear the same clothes all the time, until he came to Tyrwhitt House. He was given a bath on arrival, and fresh clothing was found for him – which he wore for another year. Fred's regimental association helped out towards his telephone bills, as we all felt that the telephone was a kind of lifeline. I suppose we were, collectively, the only friends he had in the world. And indeed he told me this more than once.

Travelling around a vast area, given a variety of urban and country addresses to visit, was not easy to begin with. I collected an assortment of street plans for the small towns and large villages, usually from an estate agent. The larger conurbations called for a street atlas which had to be bought, there was nothing else for it. The Ring Roads around London and Birmingham were a godsend. Then one had to avoid the town and city centres early morning and late afternoon. Most of my clients would not expect a visit until after 10 a.m., in any case. Finding one's way in the open countryside presented a different set of problems. The address might be an isolated farm, or otherwise vague. Sometimes there was nothing for it but to call at a garage, the village Post Office, a newsagent, or the Police Station. New streets, not on the map, presented another problem. Fire Station officers were always helpful in this regard.

Some clients were apprehensive about meeting for the first time. Thus, if I knocked on one door, certain the man was in, but no one responded, I drove away, round the next corner from where I could still see the house. I waited until he opened the door and then strolled over, as if nothing was amiss. Usually clients were pleased to see someone who could 'talk their language'.

Other clients went to more extreme measures to avoid me, by deliberately going out. They had all been sent a letter to say that I would call at an approximate time, on a given day. My letter also stressed that times could not be accurately forecast because of road conditions, the weather, or other delays. For example, I might be delayed by a client who was feeling particularly lonely and needed an extra half hour's chat. One thing was certain, I made it a point of principle always to arrive eventually on the promised day. This aspect is important, for many clients were nervous, depressed or otherwise manic, and could have become distressed if an expected visit did not materialise.

On one occasion, we listened to the weather forecast from home and the news was that Cirencester was cut off by deep snow. I rang the hotel where overnight accommodation had been booked to be informed, 'If you get this far, you'll not get away again!' So, I had to ask the Office to cancel the next day's calls by ringing round the clients. Most of them were not expecting me to attempt the journey anyway. My visits were usually arranged in a round trip of two to four days.

The worst experience I had was in South Wales. I had booked in at a motel to the west of Swansea and the weather forecast was not good. The morning after I arrived, I found myself following a double-decker bus up a steep hill. When the bus began to slide backwards on the ice, towards me, I decided that it was just not safe to continue. I struggled in the deep snow and ice back to the motel where I was to be stranded for the next three days. The central heating broke down, as did the hot water system. So I sat in the lounge next to a log fire anxiously watching the weather news on the television.

When I tried to phone my clients, I was unable to get through. To my utter amazement, I was able to raise the office in Wimbledon from where staff were able to contact my patients without undue difficulty. I never understood why this was so. Breakfast was hilarious. I said to the man who brought my bacon and eggs, 'I haven't seen you before'. 'No,' he replied, 'chef can't get in. I'm the plumber, here to repair the burst pipes!'

On another occasion, I arrived at a hotel, having booked in earlier, to find the front door locked and all the tables and chairs piled up inside. A van drew up beside me on the car park. 'Are you anything to do with this place?' asked the driver. 'They've gone bankrupt and they owe me some money!'

One client telephoned the office to acknowledge my intention of pay him a visit. 'My dear chap,' he said, 'I don't know how you manage to get round to call on people like me in isolated places.' When I arrived on his doorstep, exactly on time, at 10 a.m. I was greeted by, 'Who the Hell are you?'

As I travelled round, since 1984, Travelodges sprang up, near the motorways at first, then elsewhere. Now there are Travel Inns, Granada Lodges and others, forming a useful network of places to stay at. Advance booking is an advantage, and I found them all of a similar high standard of comfort for my purposes. Now that the Society employs a larger number of welfare officers with smaller territories, the requirement to stay away from home on a regular basis no longer applies to all of them. However, I can strongly recommend the principle of setting up in the centre of an area having a large number of clients, and staying there until all have been seen. Much time and effort is thereby saved. Evenings were often spent in writing up case notes and correspondence, and transferring the information onto micro-tapes which could be sent through the post to my Secretary. She then typed them up by the time I arrived back at the office.

Every visit was recorded, including what was said or promised to a client and how he reacted to it. In view of the poor memories of some of them, this was found to be extremely important. Any advice given was carefully logged, in the best interests of doing the job efficiently.

I have referred to those clients who tried to avoid me and, in an attempt to minimise any fruitless journeys, I always tried to write to clients in advance, about ten days ahead. This was to provide an opportunity for them to inform me if it was not convenient for me to call for any reason. Word that a patient was in hospital might lead to him being visited there. Should I receive a cancellation, it might be possible to visit someone else. My letter always mentioned this fact. If a client was simply not there, without explanation, I would give him a second or even a third chance before writing to say that if ever he needed me, he knew where to find me. Meanwhile, if appropriate, the War Pensions Officer or the GP would be informed, along with anyone else who might have asked me to visit the client.

Should a client be in line for a visit to Tyrwhitt House, the next stage was always to write to his GP, with the patient's written consent, requesting an up-to-date report on the patient's condition (in confidence) and his suitability for admission to a convalescent home. Replies were always submitted to the Society's Chief Consultant, who had the final say in who was to be admitted.

Some clients can have a vivid imagination. I received an appeal from a man in his fifties who lived in a country village with his elderly mother. It was a strange letter, referring to Service in the Army, but parts of it did not ring true. Nevertheless, I resolved to pay him a visit. His mother looked on anxiously while I asked her son all the usual questions about himself. I concluded by saying, 'From what you say, I don't think you were ever in the Army at all.' His mother said, 'There you are, I told you, you should not be wasting the Major's time.' I continued, 'If you weren't ever in the Forces, why did you write to me in the first place?' He replied, 'Well, you see, I thought I might have been!' There was no answer to that!

One day I received a formal request from the Ministry of Defence to visit a retired officer, formerly employed by the ministry at an office in London. This officer kept on turning up for work, although he was supposed to have retired from them some years previously. Each time he arrived he reported to his old department, was told there was no work for him, and he went home on the next train. When I called, he said he was still receiving his monthly pay and he felt he had to go to work in order to earn it. I asked to see some evidence of this, and found that he was drawing a Service pension monthly from the Paymaster General through his bank. I pointed out the difference between the correct pay for his old rank and his present pension, but he remained unconvinced. His memory was patchy. I asked him if he had ever received a letter of thanks from the Military Secretary, marking the end of his Service in the Army. He was not aware of one. Nor had he received a letter from the ministry at the time of his retirement, he claimed.

In the end, I asked the General Officer at the head of his department if he would care to write a letter of thanks, pointing out that there was no longer any need for him to return to work. This succeeded in finally convincing him. But when I called to make a follow-up visit quite soon afterwards, he had been placed in a nursing home as his memory had completely gone. Perhaps the thought that the Army still needed him had kept him going.

Perhaps the strangest story of all was that of a retired Captain from an infantry regiment who lived at a guest house on the South Coast. I called at the request of his War Pensions Welfare Officer as his landlady wanted him to vacate his room for a couple of weeks while it was redecorated. Because our friend thought it was a hotel, and he owned the place, he was refusing to move anywhere. When I arrived, he had deliberately gone out. His landlady said he thought he was a Major General, still in the Army, and that the Ministry of Defence owed him the difference between the pay of a Captain and a General going back to the Second World War! I called again, was still unsuccessful in meeting him, and contacted his old regiment.

The regiment sent another retired officer, disguised as a holidaymaker, and he booked in at the guest house. The 'General' duly came down to breakfast, dressed in khaki shorts and a shirt bearing the insignia of a major general on his epaulets. The holidaymaker engaged the general in conversation and he was informed, 'The Russians have invaded Scotland and the regiment are bravely holding them back. I'm in charge of the defence

of the south of England. I have set up road blocks in the New
Forest, but they will let you through if you mention my name.'
That day a confused local baker telephoned the landlady to confirm an order the 'General' had placed for two thousand loaves
of bread to feed his troops!

Earlier, the Police had escorted the Captain back to the guest-
house. He was wearing an old-fashioned Admiral's uniform and
was claiming to be in charge down at the harbour, where he was
causing chaos. *HMS Pinafore* was being put on by the local ama-
teurs at the time!

Another client once said to me, 'Excuse me, Sir, please don't sit
between me and the electric fire. I'm expecting an important
message to come through the wires!'

There are many other amusing stories one could recount, in
similar vein. In facing difficulties of this sort, the best approach
was always to humour the client, before changing the subject. If
I did not know the client well, and suspected an inclination to
violence, I tried to keep myself between the patient and the door,
just in case he should turn awkward. This hardly ever happened,
but the possibility was there more than once.

A few of my clients were little short of being religious fanatics,
and it was best not to become too involved with them. Some
wore religious insignia and would attend church on a daily basis,
if possible. One never attempted to criticise them for this. My
own view was that, if this gave the man some comfort in life,
who was I to disagree with him? The outcome tended to be calming and, if the local clergyman was sympathetic, then this could
only be good for him. Sometimes, the priest was the only friend
the client had locally, with whom to discuss his problems. The
Churches have, in recent years, tended to include mental welfare
as an aspect of training for ordinands. This can only be to the
advantage of everyone.

On one occasion I was contacted by a retired Church of
England clergyman. He was keen to visit clients on our behalf,
even though his age and physical condition led me to be wary of
accepting his offer of help. I therefore obtained his permission to
consult his GP for confirmation of his fitness to undertake this
type of work. The doctor responded with a report indicating that
there was a history of mental illness which could be worsened

should he be reminded of his own past problems. He had been retired from the Church's active ministry for this very reason. Thus, there was nothing for it but for me to see him again and, in a kindly manner, decline his offer of help. The GP promised to arrange some counselling sessions which he thought might help this well-meaning man.

My own religious faith was of great comfort to me. I would never raise this topic myself, but was happy to discourse if the client brought this issue up. We might then be able to have a helpful discussion, and there were many times when a client, near to the end of his life, was helped in this way. My own feelings towards the work I was doing were greatly helped by my own faith. This was especially the case on those days when I felt I was getting nowhere, when clients were difficult, and I began to wonder if I was doing any good to anyone at all, but I would persevere. Where there was a successful outcome it would all seem worthwhile. My religious faith and beliefs enabled me to live and fight another day.

I have now reached the end of a journey, travelling some quarter of a million miles and making nearly five thousand home visits for the Society. I would not have missed the opportunity of doing this work for anything else in the world. Of all the appointments I held, both inside and outside the Services, this was the most rewarding by far. I put all the effort of which I was capable into it and I can say, with confidence, that it was a source of great happiness for me. Looking back, my ten years as a Welfare Officer with the Ex-Services Mental Welfare Society has given me a tremendous feeling of satisfaction and of humility.

As well as paying tribute to all the staff employed by the Society, including those who worked in our homes and in the offices, and behind the scenes, I would also like to mention the large numbers of volunteers employed by the other charities who gave up their own time to bring cases to my attention. Sometimes they also followed up the work that I initiated with the clients. The members of the War Pensions Committees, again all volunteers, and the Welfare Officers employed by the War Pensions Agency, deserve special praise. Their devotion to clients is of the highest order. Without all these friends and colleagues, I could never have succeeded in this work.

And lastly, to those who should not be forgotten or discarded, to those ex-Servicemen and women who have served their country, who have suffered trauma, illness and distress as a result of Service, God Bless you and thank you for being my inspiration.

13

Bringing the Story up to Date

Chapter 8 detailed the First Gulf War which occupied our British Forces from November 1990 until the end of February 1991 when we drove Saddam Hussein's Iraqi forces out of Kuwait. Although the joint armies destroyed Saddam's armed forces, regime change was not effected.

On the home front the IRA continued to be active but there were no thoughts of suicide bombing. Their leaders were well known to MI5 and the SAS so that when I visited an SAS veteran in Hereford prior to a parade to mark the Regiment's fortieth anniversary in January 1992, I expressed some concern. SAS members are never photographed and here they were about to stage a parade to Hereford Cathedral. I feared the IRA might launch a "spectacular" show of defiance. I was told "They wouldn't dare. After all we know them all personally!" In Northern Ireland, breaches of the cease-fire, which came later, are rare nowadays.

The events of 11 September 2001, known as "9/11", were quite different from anything else the world had seen hitherto. Civilian airliners were hijacked by gunmen and flown into the twin towers of the New York World Trade Center at 1.58 and 2.16 p.m. BST. The buildings were destroyed, killing many thousands in their offices and in the streets. These attacks, and others directed on the Pentagon in Washington and Camp David, were attributed to the al-Qaeda movement, led by Osama bin Laden, believed to be in Afghanistan.

I was in East Berlin when the news broke in the late afternoon, following a day of sightseeing, part of a tour of Baltic countries. There were 1,000 American passengers on board our cruise liner. For all of us, henceforth, the world would never be the same as it was, as fears grew of renewed terrorist attacks. As we boarded our train to leave Berlin the streets were suddenly lined with German riot police, armed with machine guns, and when we arrived back at our ship there were frogmen looking for mines planted beneath the hull! Our voyage continued, with the express permission of all the countries on our route. No one could have gone back to New York as all flights in and out had been suspended. All entertainment on board was cancelled to be replaced by combined Christian/Jewish acts of worship. These served to comfort the passengers and honour the dead. There were free phone calls to America and we received expressions of support from all the countries we visited – including Russia!

It seems that because President George Bush Senior had let Saddam Hussein go free after the recapture of Kuwait, this was taken as a sign of weakness on the part of the Americans. Comparisons were made with the American retreat from Vietnam. So, when we invaded Iraq in March 1993, it was alleged by some Arabs that George Bush Jnr. was avenging his father's defeat and defending the honour of his family. The Second Gulf War was over in three weeks and Saddam's statue in Baghdad was toppled. Bush referred to Iraq as a "crusade" against terrorism. This was an unfortunate choice of words because it led Muslims to call for "Jihad", or Holy War, against the "Infidels". The argument still continues about the legitimacy of the invasion of Iraq and the change of regime there. However, terrorist attacks and foiled attacks in the UK led the British Government to join an American-led coalition to prevent the Taliban, a fanatical Muslim organisation, from re-occupying Afghanistan and training terrorists there, including a few British nationals.

With the fall of Saddam, his army, the police and the Ba'ath Party were disbanded. It was not surprising that the country descended into chaos with looting, unemployed armed soldiers, death squads and suicide bombers as Shi'ites and Sunni Muslims attacked each other.

The British Army served in Basra, Iraq's second largest city, but the atmosphere was very tense. Saddam had kept al-Qaeda out of Iraq, but now hundreds of disaffected suicide bombers arrived from all over the Moslem world. US and British forces only slowly reinstated and retrained the Iraqi Army and Police. Meanwhile al-

Qaeda was offering huge sums as a bribe to anyone who killed an American soldier.

The country became littered with IEDs (improvised explosive devices or home-made bombs). The Taliban mentality was, "If you keep up attacking them for long enough, foreign soldiers will leave the country to us." Iraq is still suffering from this situation although there are thousands of American soldiers still there, mainly in barracks. The British Army recently left Basra in order to concentrate efforts in Afghanistan where we began operations in 2001.

When we look at case histories from the most recent campaigns, the similarities with those from earlier chapters in this book are striking. The effect on the human mind of stress caused by shock, fear and despair has been seen in Iraq, just as in the World Wars or the lesser war situations which followed.

David was a client brought to my attention by Combat Stress in one of their fund-raising endeavours. He had served in Kosovo which in itself was no 'picnic'. By 2003, David was in Iraq as a Section Commander in Basra. A loud explosion was caused by two Challenger tanks being fired upon and David, a trained 'medic', went to the scene to treat survivors. One tank was burning and two crewmen were missing. David stumbled on their burning bodies. That sight was to haunt him long afterwards (see the case of Brian on page 44). The memory remained with him when he returned to duty in Germany and where he was told his problems were insignificant. After leaving the Army it was not until Combat Stress took him in to an ex-Service atmosphere of care and understanding that he began to recover.

I was reminded of another man referred to me by his GP in the 1980s. He was having nightmares about going back to a blazing tank after the Battle of El Alamein where he was retrieving bodies. His wife was so grateful when Combat Stress took him to Tyrwhitt House and I helped him obtain a War Disability Pension.

Herman was an officer who rose to the rank of Captain. He encountered severe stress every time they sent him to Iraq, in 2003, 2004 and 2006. The third posting was just too much for him and he was medically discharged with mental illness. Herman was a medical administrator and back in 1999 he had been dealing with casualties in Kosovo. By 2003 he was in Shaibah, Iraq, in control of casualty evacuation. A year later in the field hospital in Basra Herman was doing the work of more than one officer and sleeping no more than 10 or 12 hours a week for weeks on end. At that time the army was still suffering two fatalities a week out there with many times this number of seriously injured personnel.

When Herman finally broke down in 2006, he was shunned by others who should have supported him. This made life even more stressful and made him feel much worse. Subsequently he has received support from "Help for Heroes" and it was they who introduced him to me. He had read the first edition of this book and now intends to make contact with Combat Stress when he leaves the Army in 2010.

Repetition of stressful situations occurred in the First World War when men were repeatedly made to go "over the top" or be shot for cowardice. For example, RAF aircrew who refused to go on more operations than they could withstand were said to be of "Low moral Fibre" (see p. 78).

Long after the British Army had left Basra, Iraq was still in a dangerous state. There were terror attacks inside the "Green zone" in Baghdad where it was supposed to be well defended. As I write, in 2009, a co-ordinated series of huge explosions have rocked the city at a time when the Iraqis have been promised fresh elections. Al-Qaeda has been blamed for these attacks, obviously planned, and questions are being posed as to how they were allowed to take place.

Afghanistan has always been a lawless place. Queen Victoria's Army never succeeded in making a lasting peace as the tribesmen on the North West Frontier were a law unto themselves. November 2001 saw the start of a military mission to stabilise and pacify Afghanistan. Progress has been understandably slow.

The Parachute Regiment arrived in 2006. It was reported that the Taliban controlled even the small towns. There were mines everywhere, some left by the Russians who had failed to subdue the country, leaving behind huge quantities of wrecked military hardware.

A typical scenario was as follows.

Casualties were evacuated by Black Hawk helicopter, some-times after a long delay. The Taliban were waiting for them and attacked our wounded men. One had his legs blown off, another bled to death while waiting to be rescued. The Taliban preferred rocket-propelled grenades (RPGs) with which they peppered our helicopters with shrapnel. In one day, the Third Parachute Regiment Battle Group suffered casualties, evacuated to Camp Bastion (the main military base in Afghanistan where there is a military hospital capable of the most advanced surgical procedures) from three locations. Two men died on the way, three had legs amputated – all in the space of fourteen hours. This was 6 September 2006.

At this time the Taliban were terrorising or murdering anyone associated with Government or Coalition forces. Dependence on helicopters was shown to be uncertain. Local Afghan police were unreliable while some, we knew, were spying for the Taliban who paid better wages. The enemy often seemed to know about our troop movements. Being under continual attack caused the men's mental stamina to run down, at a time when physical conditions were very demanding. It was not unheard of for the food to run short in a forward base, and there was always a fear of becoming a casualty and thus letting down one's comrades.

Being under siege in some locations could be very wearing. Yet, when it was discovered that the Taliban were having to bring in recruits from Pakistan and elsewhere beyond Afghanistan, this was a sign that we were perhaps winning. We knew some of the rebels were strangers by the manner of their speech. We never knew how many casualties we inflicted on them because the Taliban always took away their dead and wounded – sometimes after dark.

When the Paras lost one of their NCOs, the men became very despondent; it could have been any one of them. He sacrificed his life for his friends by moving forward into the enemy fire and protecting them. For this outstanding act of gallantry he was awarded the Victoria Cross.

For soldiers, conventional fighting consists of going on the attack followed by a lull or a break while the position is consolidated. Afghanistan today consists of constant stress and trauma for those who serve there on operational duties. Some of the men, understandably, questioned why they were there; the events of "9/11" were American, not British. In Iraq there were roadside bombs going off, followed by some respite. In Afghanistan you could be under continuous attack or sudden bursts of fire at any time of day, or night. It was said that the Taliban don't attack after dark – but they certainly do!

In a town called Musa Qaleh the local people lived in fear; many closed their shops and left. At a meeting, they asked the British to leave and the Taliban to stop fighting. The town was in ruins and a cease-fire was agreed. The Afghan flag was hoisted and the locals gave food to the troops as a gesture of friendship. The British Army forces were not driven out but the Taliban broke the truce by murdering the leader of the local elders.

The Parachute Regiment never did much by way of reconstruction, they were too busy fighting but they did assist in saving the hydro-electric dam at Kajak, which supplied power for a large area – a natural target for terrorism.

Months afterwards, cases of PTSD were reported, originating in flashbacks to the faces of dead comrades in battle. Combat Stress told me of a Lance-Corporal on a Scimitar armoured reconnaissance vehicle. In 2003, Derek was in Al Amarah, a very unstable part of Iraq. When he reached Baghdad he realised he was not himself when he developed OCD (see p. 239). Derek became obsessive about his gear, a condition brought about by sheer exhaustion. Back in Iraq in 2005, it was even more stressful for Derek. The place was unfriendly, lawless and there were roadside bombs. He was in charge of some of the convoys and the men looked to him for a good example. In October 2006 Derek found himself in Afghanistan. By then he was having feelings of anger and confusion. On arrival at their new base they were fitting out vehicles until midnight.

One day, a Scimitar had a fuel problem: while under mortar fire the engine was stalling with local people watching nearby. The soldiers were in that village, under fire, for over two hours. It was after this that Derek was sent home to the UK feeling guilty at having to share an aircraft with physical casualties. Back home Derek broke down and tried to kill himself. Then someone told his story to Combat Stress and he was taken to Audley Court where he was able to regain his confidence. Derek says the Society saved his life and he is now looking forward to a new career.

Mark Omrod is a Royal Marine who landed at Kandahar in October 2007. Earlier he had served in Iraq in 2003 as a field

Ambulance driver. It was a job where the atmosphere was comparatively quiet but they all wore NBC suits (p. 181) and there were threats from Scud missiles.

Afghanistan was to be completely different. His book *Man Down* is a first-class personal record of a stressful time culminating in the loss of both his legs and one arm. He stepped on an IED whilst on patrol on Christmas Eve 2007. Mark was in FOB (Forward Operational Base) Robinson in Helmand Province. He had been taken there in a party of reinforcements in a Chinook helicopter from Camp Bastion.

At Robinson they might take hits from a Chinese-made 107mm rocket which could inflict severe damage while the men took refuge in a shelter below ground. When the army returned fire soldiers were instructed never to endanger the local civilians. But this was not advice respected by the Taliban. There was frustration amongst the soldiers who could see shapes of what looked like Taliban fighters only to be told by HQ not to open fire albeit that it might have silenced the RPGs (Rocket Propelled Grenades). There were always dangers in calling for air support which might not be of great accuracy.

Mark Omrod's company had not been out on patrol for a month. Having recently been reinforced, those in charge thought that the British contingent ought to show its strength. The bomb which almost killed Marine Omrod made a crater 8 feet deep and 15 feet across. He was thrown into the air and blamed himself for what happened as they were not even under fire at that time. Mark's first reaction was to ask his nearest comrade to finish him off as he had no future in the Marines or anywhere else.

Luckily, the extreme shock of the explosion caused his blood vessels to contract just enough to keep him alive. An anaesthetist in the back of the Chinook helicopter which took him to Camp Bastion thought Mark was already dead but they continued to pump fluids into him and his breathing returned.

Back in Robinson the officers and NCOs did their best to give the men some Christmas cheer while morale had sunk to a low ebb. On the credit side, Mark's face was unscathed, although his girlfriend feared some brain injury at first.

On arrival at Selly Oak Hospital in Birmingham, Mark's legs and arm were amputated further to help in the future use of artificial limbs. Before going into theatre he asked his girl if she would marry him. Before he passed out she said she would accept Mark as he was.

It was a slow recovery coming to terms with the loss of limbs and coming off drugs. Then there was the risk of infection. But his Royal Marine training brought him through; he showed determination, courage, cheerfulness in adversity and unselfishness. Mark needed all of this if he was not to feel sorry for himself. Royal Marines do not indulge in self-pity. These aspects of Royal Marine training helped Mark to survive whereas the "stiff upper lip" attitude does not suit everyone, depending on their mental state, as we know.

As soon as possible, Mark was able to get himself into a wheelchair, unaided. His fiancée and relatives were able to stay in accommodation for military families at Selly Oak.

When his condition had stabilised sufficiently, after six weeks, Mark was transferred to the Defence Medical Rehabilitation Centre at Headley Court, Leatherhead, where they supplied the best prosthetic limbs in the world. They told him he would walk again within a year!

At Headley Court, Mark was visited by Mick Bremner, a Sergeant in the Royal Signals, blown up by a suicide bomber in Baghdad in 2004. Although Mick lost both legs, he was still in the Army, training as a paralympic oarsman under Matthew Pinsent. Three years on, Mick was walking normally. He showed Mark limbs with no straps, worth £20,000 each and powered by batteries controlling movement through computers.

Headley Court looks after all aspects of amputation. It is a military establishment but the charity "Help for Heroes" gave them £8 million for a swimming pool. Princes William and Harry have visited Headley Court, later raising money for "Help for Heroes" and SSAFA. SSAFA opened new accommodation called Norton House near Headley Court, costing £2 million. Mark was able to stay there with his family when they visited him.

Mark Omrod did a parachute jump and raised £3,000 for SSAFA through sponsorship. He is now back at 42 Commando Royal Marines at Plymouth. They have provided an officer's married quarter specially adapted with lifts and ramps. He lives with his wife and is employed as a clerk dealing with computers.

Help for Heroes has donated large sums towards facilities for Combat Stress and Headley Court, both near Leatherhead. Modern surgery is saving the lives of many who, as recently as in the Falklands, would have died. Now we have considerable numbers of soldiers learning to make new lives, having lost one or more limbs.

SSAFA has funded two houses at Selly Oak and Headley Court to accommodate families. In December 2009 Max Hastings of the *Daily Mail* asked readers to donate money to support his appeal. Earlier in 2009 the *Sun* newspaper publicly "adopted" Help for Heroes, asking readers to donate money, time and talents. There has been a huge response. Disabled personnel are also going to other rehabilitation units, elsewhere in the country.

Injured forces personnel prefer to be part of a military unit but there are few opportunities for this. Mark Omrod was fortunate in having a chance to remain with his own unit and to continue to serve. The civilian world is not always welcoming or understanding. However, at Headley Court they have around 200 patients, all hoping one day to be discharged.

Some patients have psychological problems which can only diminish their employment prospects (see Chapter 11). This applies not only to those with PTSD. Some soldiers are in denial about their situation and they feel defiantly negative about life. These were amongst the most difficult cases with which I ever had to deal. Some today cannot bear to look at their injuries. Some never did have much by way of job prospects outside the Army. What can they do now? Parents and partners suffer and there is a sense of guilt related to the Iraq and Afghanistan wars of debatable origins.

The medical facilities at Headley Court are state of the art but facilities elsewhere are lacking. Soldiers now receive compensation payments similar to civilian injury settlements, but they need help in using their money wisely, assuming they may never work again. Our injured ex-Service personnel don't want pity but they do need our respect. There is an idea of using them as support staff with the Police and other uniformed civilian services where there is a structure of comradeship to which they might relate.

Combat Stress (or the ex-Services Mental Welfare Society) is busier than ever it was in my time. Future prospects involve playing a vital part in a recently announced Community Veterans' Mental Health Service.

One of my objectives in writing this book, now in its second edition, is to bring to a wider public, knowledge of the excellent and invaluable work of Combat Stress. When I worked for the Society full-time, I met no end of people for whom my visit was the first time they had ever heard of what we did. Combat Stress tell us that in 2007 they had 1,200 new cases and demand for their services and engagement has risen by 53% over the past three years. The average age of new cases is 43 and, on average, clients had waited 13.6 years after discharge before coming to Combat Stress. The Society has increased its support staff from 180 in 1998 to 250 in 2009.

Help for Heroes, led by their Chief Executive, Bryn Parry, has agreed to donate £3.5 million in order to extend facilities at Combat Stress, Leatherhead. This will facilitate taking in more patients as well as accommodating their carers. By the summer of 2009 there were 4,000 cases on the Society's books and, of these, 1,257 had newly arrived during 2008–9. The summer of 2009 also saw an extension to the Activities Centre at Tyrwhitt House.

In the early days, before my time (I started in 1984), the Society would treat anyone who was ex-Service and who was regarded as having a mental condition related to military service. During my time it became necessary to concentrate our efforts mainly on those who had some financial support if they stayed for treatment in one of our homes.

The award of a War Disability Pension could carry a periodic entitlement to a treatment allowance, with expenses paid. I therefore worked all the harder to help my clients, without a War Pension, to apply for one. Others might receive financial help from their appropriate Service charity. Results were patchy, however.

It has recently been announced, officially, that Combat Stress treatment for patients will be funded by the NHS and not by the Ministry of Defence. Local Primary Healthcare Trusts will fund treatment for mental health care, and the Society will be competing with other statutory and voluntary bodies for the funds the Government makes available.

Unlike the War Pensions Scheme, which it replaced, the Armed

Forces Compensation Scheme does not pay for healthcare. Combat Stress will employ Community Outreach workers, working with our Regional Welfare Officers, who now number 16. These Welfare Officers are shared between the Society's three homes in Leatherhead, Shropshire and Ayrshire. In this age of rapid communications I imagine they will work from home, to an extent. In my day, there were only six Regional Welfare Officers covering the whole of the British Isles.

Appendix C contains details of six NHS Mental Health Pilot Schemes, announced on 3 December 2009. The broad scheme recognises that the NHS can no longer easily access expertise in military mental health matters. This is for two main reasons: the number of Service casualties is far greater than hitherto; secondly, the number of medical and nursing personnel with Service experience continues to decline. The pilot schemes will last for two years, to be evaluated before further thought is given to these problems.

The new community-based trial model has been developed following consultation between the MoD, the UK Health Departments and Combat Stress. It is now recognised, officially, that ex-Service folk feel that civilians have little conception of Service life. Veterans can be reluctant to seek help and delay can make its potential success all the more difficult. At the heart of the new system is a Community Veterans' Mental Health Therapist. These people do exist nowadays, but on leaving the Forces they have tended to be part of the civilian NHS where their Forces experience is largely unused.

The therapist can refer cases for treatment at a Combat Stress home. It has been officially stated that the new clinics will be open to all ex-Service personnel requiring diagnosis and treatment for anxiety, depression, alcohol and drug abuse, and PTSD.

Individuals will access the programme through self-referral, Social Services departments, the ex-Service organisations, the Veterans' Welfare Service, the individual's GP, or via Combat Stress and its Regional Welfare Officers.

It is now ten years since this book was first published, but during that decade interest in the psychological problems associated with military service has increased beyond all our expectations.

*   *   *

Appendix C has been completely revised and it is now up-to-date at the time of compilation – December 2009.

Looking back to chapter 2, 'The Early Activities of Ex-Service Charitable Organisations', we find the early work of SSAFA, supporting Service families left at home during the South African War. Page 17 shows how charitable work with the poor and the unemployed developed in Edwardian England. The original "Combat Stress" was established in 1919.

All this much needed work was entirely supported by voluntary activities and subscriptions.

On 12 March 2010, in order to finance its share in the new Community Veterans' Mental Health Service (see p. 307), Combat Stress launched a major national appeal for £30 million. This appeal is supported by the Prince of Wales and Baroness Thatcher. The Lottery Fund has also promised a £35 million 'Forces in Mind' Programme of help for disabled ex-Service Personnel.

Thus the wheel has gone full circle and we are back to a situation where the ex-Service fraternity are expecting most of their much needed support to come from the country at large.

What has changed?

At the end of the Second World War the Welfare State, including the NHS, was born. So it was that anyone leaving the Forces, having been disabled, felt they had a claim against the Government.

This is an attitude of mind urgently in need of revision, and Combat Stress is right to ask the whole country to support those who suffered, and are going to suffer, after serving this country so well. The national appeal is the way forward and I support it wholeheartedly.

Author's note:

The First Edition of this book was produced in 1998 with the written permission of the Director of Combat Stress. I still hold copies of his authorisation. All Royalties I received were handed over to the Society. The entire production costs of both editions have been met by the Author himself.

For any matters raised in the book I can be contacted on: 01202 680160. E-mail: barballen@blueyonder.co.uk

Select Bibliography

Ahrenfelt, R. H. *Psychiatry in the British Army in the Second World War*, Routledge & Kegan Paul, 1958.

Ashcroft, James, *Escape from Baghdad*, Virgin Books, 2009.

Baker, Dennis. *Soldiering On*, Sphere Books, 1983.

Bishop, Patrick, *3 Para – Afghanistan 2006*, HarperCollins, 2007.

Bootle, William. *I Wanted to Fly So I Joined the RAF*, Amberwood Graphics, Swanage, Dorset.

Clark, R. W. *Great Moments in Battle*, Phoenix House, 1960.

Coleman, Vernon. *Life without Tranquilisers*, Piatkus, 1985.

De la Billiere, Peter. *Storm Command*, HarperCollins, 1992.

Fergusson, Bernard. *Beyond the Chindwin*, Fontana, 1957.

Garrett, Richard. *Prisoner of War: The Uncivil Face of War*, David & Charles, Newton Abbot, Devon.

Handbook of Army Health, War Office Code No. 5691-1 1950.

Haslam, Michael. *Psychiatry Made Simple*, second edition, Butterworth-Heinemann, 1993.

Hodson, J. L. *Gentlemen of Dunkirk*, Cherry Tree, 1940.

McLaughlan, Redmond. *The Royal Army Medical Corps*, Leo Cooper, 1971.

McManners, Hugh. *The Scars of War*, HarperCollins, 1993.

Milligan, Spike and Clare, Anthony. *Depression and How to Survive It*, Arrow Books, 1994.

Moore, William. *The Thin Yellow Line*, Leo Cooper, 1974.

Murphy, Elaine. *Dementia and Mental Illness in Older People*, Papermac, 1993.

Muss, David. *The Trauma Trap*, Doubleday, 1991.

Omrod, Mark, *Man Down*, Bantam Press, 2009.

Rapoport, Judith. *Obsessive Compulsive Disorder: The Boy who Couldn't Stop Washing*, Collins, 1990.

Report of the War Office Committee of Enquiry into Shell Shock, HMSO 1922.

Sargent, William. *The Battle for the Mind*, Heinemann, 1957.

Silverstone and Turner. *Drug Treatment in Psychiatry*, Routledge & Kegan Paul, 1978.

Skinner, Harold A. *Guest of the Imperial Japanese Army 1942–45*, Chaffer & Son, Chichester, 1993.

Stewart, Bob. *Broken Lives*, HarperCollins, 1993.

Vaughan, Geoffrey D. *The Way it Really Was*, Granary Press, Budleigh Salterton, 1985.

Walker, Ernest. *The Price of Surrender*, Blandford Books, 1992.

Weekes, Claire. *The Latest Help for your Nerves*, Angus & Robertson, 1989.

Whitely, J. S. and Gordon, John. *Group Approaches in Psychiatry*, Routledge & Kegan Paul, 1979.

Reference Works:

Charities Digest, Published annually by Family Welfare Association, 501, Kingsland Road, Dalston, London E8 4AU.

Charity Choice – The Encyclopaedia of Charities (10th edition, 1998), Waterlow Information Services.

The Disability Rights Handbook, Published annually by The Disability Alliance, 1st Floor, East Universal House, 88–94 Wentworth Street, London E1 7SA

Hart, Sir Basil Liddell (ed.) *History of the Second World War*, Macdonald & Co., 1989.

National Directory of Hospitals and Homes Providing Rehabilitation, Short-Term Care and Convalescence 1991 edition. Published by Independent Healthcare Association, 22 Little Russell Street, WC1A 2HT.

Pimlott, John (ed.) *British Military Operations 1945–84*, Hamlyn, 1984.

Additional material provided by:

Ex-Services Mental Welfare Society (Combat Stress)

The 'Not Forgotten' Association

The Royal British Legion

SSAFA Forces Help

War Pensions Agency, Norcross

Appendix A   Military and other Terms and Abbreviations

ACI	Army Council Instruction.
AFC	Air Force Cross.
ARP	Air Raid Precautions.
ATS	Auxiliary Territorial Service (later Women's Royal Army Corps).
Auschwitz	German Concentration Camp.
Axis	Alliance of Germany, Italy and, later, Japan.
BEF	British Expeditionary Force (1914 and 1939).
BAOR	British Army of the Rhine.
BLESMA	British Limbless ex-Servicemen's Association.
Bren Gun	Light Machine Gun (LMG) developed in the 1930s.
BRU	Battle Recovery Unit.
Chindits	British force operating behind enemy lines in the Burma campaign.
CO	Commanding Officer.
CRU	Civilian Resettlement Unit.
CT	Communist Terrorist (Malaya).
DCI	Defence Council Instruction.
D Day	Date of Allied landings in Normandy, 6 June 1944.
DFC	Distinguished Flying Cross.
DFM	Distinguished Flying Medal.
DHSS	Department of Health and Social Security.
DRO	Disablement Resettlement Officer.
DSS	Department of Social Security.

EMS	Emergency Medical Service.
EOKA	Cypriot terrorist organisation.
ESMWS	Ex-Services Mental Welfare Society.
FHS	Forces Help Society.
FEPOW	Far East Prisoner of War.
FOB	Forward Operational Base
GCHQ	Government Communications Headquarters.
GI	General Infantryman (United States Army).
IED	Improvised Explosive Device
LMF	Low Moral Fibre.
Mau Mau	Kenyan secret terrorist organisation.
Military Secretary	Officer responsible for the administration of officers in the British Army.
MoD	Ministry of Defence.
MPAJA	Malayan People's Anti-Japanese Army.
NBC suits	Protective clothing issued against Nuclear Biological and Chemical warfare.
NCO	Non-commissioned Officer.
NFA	The Not Forgotten Association.
OAP	Old age pensioner.
PBI	Poor Bloody Infantry (World War I)
PLO	Palestine Liberation Organisation
POW	Prisoner of War.
QARANC	Queen Alexandra's Royal Army Nursing Corps.
RAEC	Royal Army Educational Corps.
RAFBF	Royal Air Forces Benevolent Fund.
RAMC	Royal Army Medical Corps.
RAOC	Royal Army Ordnance Corps.
RASC	Royal Army Service Corps.
RBL	Royal British Legion.
RCT	Royal Corps of Transport.
REME	Royal Electrical and Mechanical Engineers.
RFEA	Regular Forces Employment Association.
RNBT	Royal Naval Benevolent Trust.
RNH	Royal Naval Hospital.
RPC	Royal Pioneer Corps.
RPG	Rocket-propelled grenade

RTU	Return to Unit.
SAS	Special Air Services.
South Atlantic Fund	Trust fund set up after the Falklands Campaign.
SSAFA	Soldiers Sailors and Airmens Families Association.
TA	Territorial Army.
Triple "A"	Anti-Aircraft Attack.
VE Day	8 May 1945.
VJ Day	15 August 1945.
WAAF	Women's Auxiliary Air Force.
WRAC	Women's Royal Army Corps.
WRNS	Women's Royal Naval Service.
WRVS	Women's Royal Voluntary Service.

Appendix B Medical Terms and Abbreviations

Affective disorder	A serious change in the emotional state.
Agoraphobia	Fear of open spaces.
Alzheimer's Disease	One of the presenile (or senile) dementias.
Benzodiazepines	A group of drugs effective in controlling anxiety states (Librium, Valium, Diazepam etc.).
Bipolar Affective Disorder	A psychotic illness where both depressive and manic swings of mood occur at different times.
Chiropractic	The treatment and rehabilitation of pain syndromes and dysfunction of the locomotive systems by the manipulation of the dysfunctional joints.
Chorea	A disturbance in the brain producing jerky movements of the limbs.
Claustrophobia	An abnormal fear of being in a confined space.
CPN	Community Psychiatric Nurse.
Delirium	An acute disorder of memory and orientation.
Dementia	An illness resulting from degeneration of the brain cells causing disturbance of memory, orientation and social awareness.
Depression	A state of dejection often accompanied with feelings of sadness irritability or anxiety.

Dissociative State	A condition caused by a long period of stress where the mind dissociates itself from the conscious environment.
ECT	Electroconvulsive therapy.
EEG	Electroencephalogram. A technique used for recording electrical activity in the brain.
Encephalitis	Brain infection caused by a virus.
EMI	Elderly Mentally Ill.
ESMI	Elderly Severely Mentally Ill.
Epilepsy	A condition where the nerve cells in the brain become hypersensitive inducing a seizure or fit.
Heminevrin	Substance used in the treatment of alcohol addiction.
Huntington's Chorea	Hereditary chorea occurring in adults, associated with progressive mental deterioration.
Hysteria	A neurotic state where a disordered bodily function exists without appropriate physical cause.
IES	Impact of Events Scale used in 'Rewind' technique for treatment of PTSD.
Jacob Kreutsfeld's Disease	A form of presenile dementia.
Leptospirosis	A severe infection of the liver which can lead to meningitis or inflammation of the membranes of the brain.
Leucotomy	A brain operation whereby certain nerve tracts are divided.
Mania	A psychiatric illness associated with an abnormally elated mood, can be seen as part of a bipolar affective psychosis.
Manic depressive psychosis	Old term for a bipolar affective disorder.
MAOI	Mono-amine oxidase inhibitors A group of anti-depressant drugs.
MENCAP	Royal Society for the Mentally Handicapped
MIND	National Association for Mental Health.

Mono-amines	A group of chemical transmitter substances in the brain.
Neurosis (Psychoneurosis)	A mental disorder, usually mild, displaying anxiety but where the patient retains some insight or understanding of his condition.
NLP	Neurolinguistic Programming. An aspect of some treatment for PTSD.
OCD	Obsessive Compulsive Disorder.
Paranoia	A psychosis characterised by symptoms of unshakeable delusion.
Paranoid	Morbidly suspicious thinking which may include persecution and delusion.
Parkinson's Disease	A progressive disease due to degeneration of the brain and which displays charcteristics of tremor, abnormal gait and slowness of movement.
Phenothiazenes	A group of chemicals used in the treatment of schizophrenia and other psychotic ilnesses.
Placebo	An inert substance, which is not a drug, which can be given instead of medication, for psychological benefit only.
Psychiatrist	A doctor who specialises in conditions which cause disturbances of mood, thinking or behaviour.
Psychoanalyst	A person, not necessarily medically qualified, who has undergone training in Freudian analytical techniques.
Psychologist	One who studies mental mechanisms, not medically qualified but may work in association with psychiatrists.
Psychopathy	A condition where the patient is persistently aggressive or irresponsible.
Psychosis	A serious mental illness, in which insight is frequently lost.
Psychosomatic	A disease where there is obvious physical change in the body but where there are psychological factors in its cause.

Schizophrenia	A psychotic group of illnesses characterised by disorders of thinking.
Spondylitis	A condition of the spine which can cause immobility.
Sympathetic Nervous System	That part of the autonomic nervous system which is concerned with preparing the subject for fight or flight.
TCA	Tricyclic anti-depressant drugs.
Tinnitus	Sensation of persistent ringing noises in the ear.
Trauma	Damage to the physical or mental processes of the patient. A severe emotional shock.
Unipolar Affective Disorder	A psychotic depressive state, endogenous depression where no phases of manic behaviour occur.
WHO	World Health Organisation.

Appendix C   Useful Addresses – Where to go for Help

Readers are advised to consult a local telephone directory or attend a library for local contacts. The information provided below is not an exhaustive list but may be a starting point for those seeking help or mutual support.

On 3 December 2009 the Veterans' Agency announced a Community Veterans' Mental Health Service. The four UK health departments, the Ministry of Defence (MoD) and Combat Stress have been working to develop and pilot a new model of community-based mental health care. It will thus become easier for veterans concerned about mental health problems to find help. The service will be led by a mental health therapist offering particular understanding of the issues facing those who have served in the Armed Forces.

Treatment options will include the use of services offered by Combat Stress. The first pilot scheme has begun in Stafford. Pilot schemes have also been launched in Camden & Islington, Cardiff and Bishop Auckland, to be followed by Cornwall and Scotland.

Meanwhile, the MoD's Medical Assessment Programme (MAP) offers expert mental health assessments to any veteran who has served in operations since 1982, living outside the areas of the pilot schemes.

COMMUNITY-BASED MENTAL HEALTH CARE

South Staffordshire and Shropshire Healthcare NHS Foundation Trust

Coton House
St Georges Hospital Site
Corporation Street
Stafford ST16 3AG

Tel.: 01785 257888 Ext 5280

Camden & Islinglton Mental Health and Social Care Trust

Graham Fawcett – CV MHT

The Traumatic Stress Clinic

73 Charlotte Street
London W1T 4PL

Tel.: 020 7530 3666
E-mail: veterans@candi.nhs.uk

Cardiff and Vale NHS Trust

Neil Kitchiner – CV MHT
University Hospital of Wales
Heath Park
Cardiff CF14 4XW

Tel: 029 2074 2284
E-mail: neil.kitchiner@cardiffandvale.wales.nhs.uk
Website: http://www.veterans-mhs-cvct.org/

Community Veterans Mental Health Service

Trevillis House
Lodge Hill
Liskeard
Cornwall PL14 4NE

Tel.: 01579 335226
Fax: 01579 335245
E-mail: Veteran.Assistance@cornwall.nhs.uk

Medical Assessment Programme – MAP

Dr Ian Palmer, Head of Medical Assessment Programme
Baird Medical Centre
Gassiott House
St Thomas Hospital
Lambeth Palace Road
London SE1 7EH

E-mail: map@gstt.nhs.uk
Freephone Helpline: 0800 169 5401

Tees, Esk and Wear Valleys NHS Foundation Trust

Psychological Therapy Service
Symon Day – Veterans Mental Health Therapist
St Aidans House
St Aidans Walk
Bishop Auckland
County Durham DL14 6SA

Tel: 01388 646802
E-mail: symon.day@TEWV.nhs.uk

Veterans First Point

Charlotte House
2 South Charlotte Street
Edinburgh EH2 4AW

Tel: 0131 2209920

THE MAIN EX-SERVICE ORGANISATIONS

Combat Stress

The Chief Executive
Combat Stress, Tyrwhitt House, Oaklawn Road
Leatherhead, Surrey KT22 0BX
E-mail: contactus@combatstress.org.uk
http://www.combatstress.org.uk

Tel.: 01372 841600

Regional Offices

Audley Court address: Combat Stress, Audley Court, Audley Avenue, Newport
Shropshire, TF10 7BP

Tel.: 01952 822700
Fax: 01952 822701

Hollybush House address: Combat Stress, Hollybush House, Hollybush by Ayr
Ayrshire, KA6 7EA

Tel.: 01292 561300
Fax: 01292 561301

The 'Not Forgotten' Association for the Ex-Service Disabled

4th Floor
2 Grosvenor Gardens
London SW1W 0DH

Tel. : 020 7730 2400
Tel. : 020 7730 3660
Fax : 020 7730 0020
E-mail: director@nfassociation.org
http://www.nfassociation.org

SSAFA Forces Help

19 Queen Elizabeth Street
London, SE1 2LP

Tel.: 0845 1300 975

Confidential Support Line: A telephone support line is available 365 days a year and provides a service which is outside the chain of command. The line is open from 10.30 a.m.–10.30 p.m. Free phone lines operate from Germany, Cyprus and the UK:

From the UK (Main Line): 0800 731 4880
From Germany: 0800 1827 395
From Cyprus: 800 91065
From the Falkland Islands # 6111
From anywhere in the world (Call-back) +44 (0)1980 630854

From Operational Theatres, to enable access through Paradigm's phone system, dial the appropriate access number then enter *201 at the PIN prompt.

Service Personnel who are absent without leave (AWOL) can now speak in complete confidence to the SSAFA AWOL Support Line Worker: Tel.: 01380 738137, 9.00–10.00 a.m. Monday to Friday (answering machine for call back outside these times).
http://www.ssafa.org.uk

The Royal British Legion

Head Office
199 Borough High Street
London SE1 1AA

Tel.: 020 3207 2100
For all general enquiries please contact Legionline on
08457 725 725
Open 10 a.m.–4 p.m. Monday to Friday, all calls charged at local rates
http://www.britishlegion.org.uk

War Disablement Pensions & War Widow or Widower's Pension

Pensions & Benefits Department
The Royal British Legion
Haig House, 199 Borough High Street
London, SE1 1AA

Tel.: +44 (0)20 3207 2164 / 2168 / 2169

Disability

Telephone +44 (0)20 3207 2158 / 2160 / 2162
Fax: +44 (0)20 3207 2218
Pensions & Disability Department
The Royal British Legion
Haig House, 199 Borough High Street
London, SE1 1AA

The Royal British Legion Village

Aylesford
Kent, ME20 7NZ

Tel./Fax Number: +44 (0)1732-870102 (answerphone during day)
http://rblvillage.legionbranches.net

The Veterans Agency (previously the War Pensions Agency) Free Veterans Helpline

Freephone 0800 169 22 77 (UK only)

BT Mobile, Orange, Virgin or 3 mobile, Customers will not be charged for the call by their provider
0800 169 34 58 Textphone facility (UK only)
+44 1253 866043 (Overseas)
Telephone lines are open: 8.15 a.m. to 5.15 p.m. Monday to Thursday; and 8.15 a.m. to 4.30 p.m. Friday

Service Personnel and Veterans Agency

Norcross, Thornton Cleveleys
Lancashire, FY5 3WP

E-mail: veterans.help@spva.gsi.gov.uk
http://www.veterans-uk.info/

Regular Forces Employment Association

RFEA Limited
First Floor
Mountbarrow House
6-20 Elizabeth Street
London, SW1W 9RB

Tel.: +44 (0)121 236 0058
http://www.rfea.org.uk/

War Widows Association of Great Britain

c/o 199 Borough High Street
London, SE1 1AA

Tel.: 0845 2412 189
E-mail: info@warwidowsassociation.org.uk
Website: http://www.warwidowsassociation.org.uk/

Help for Heroes
Unit 6
Aspire Business Centre
Ordnance Road
Tidworth
Hants, SP9 7QD

Tel.: 0845 673 1760 or 01980 846 459

Department of Veterans Affairs (Australia)
300 Latrobe Street
Melbourne, Victoria 3000
Australia

Postal: PO Box 9998, Melbourne, Victoria 3001, Australia

EX-SERVICE CHARITABLE FUNDS

ROYAL NAVY & ROYAL MARINES

QARNNS Trust Fund
Cdr Thain-Smith
Medical Division
Mail Point 3.2
Navy Command HQ
Leach Building
Whale Island
Portsmouth, PO2 8BY

Or contact Commander Julie Thain-Smith on 02392 625343
http://www.qarnns.co.uk/qarnns-trust-fund.asp

Royal Marines Benevolent Fund
RM Corps Secretariat
HMS Excellent
Whale Island
Portsmouth
Hants, P02 8ER

Tel.: 02392 547201
E-mail: royalmarines.charities@charity.vfree.com

Royal Naval Association

Room 209, Semaphore Tower
PP70
HM Naval Base
Portsmouth PO1 3LT

Reception: 02392 723823
Fax: 02392 723371
http://www.royal-naval-association.co.uk

The RN Benevolent Society for Officers

The Secretary, RNBSO
70 Porchester Terrace
London, W2 3TP

E-mail: RNBSO@lineone.net
Tel.: 020 7402 5231
Fax: 020 7402 5533
Website: The RN Benevolent Society for Officers, incorporating
   The ARNO Charitable Trust

The Royal Naval Benevolent Trust

Castaway House
11 Twyford Avenue
Portsmouth, PO2 8PE

Tel.: 023 9269 0112 (Administration), 023 9266 0296 &
   023 9272 5841 (Grants)
Fax: 023 9266 0852
E-mail: rnbt@rnbt.org.uk

Royal Naval Reserve (V) Benevolent Fund

The Cottage,
St Hilary,
Cowbridge,
Vale of Glamorgan, CF71 7DP

Tel.: 01446 771108

Sailors' Families' Society
Cottingham Road (CC)
Newland, Hull
Kingston upon Hull, HU6 7RJ
Tel.: 01482 342331

King George's Fund for Sailors
8 Hatherley Street (CC),
London, SW1P 2YY

Tel.: 020 7932 0000

Women's Royal Naval Service Benevolent Trust
WRNS BT
311 Twyford Ave
Portsmouth, PO2 8PE

Tel.: 023 9265 5301

The White Ensign Association Limited
HMS BELFAST
Tooley Street
London, SE1 2JH

Tel.: 020 7407 8658
E-mail: office@whiteensign.co.uk
Website: http://www.whiteensign.co.uk/

The RN & RM Children's Fund
Castaway House
311 Twyford Ave
Portsmouth, PO2 8RN

Tel.: 023 92639534.
Website: http://rnrmchildrensfund.org.uk/
E-mail: rnchildren@btconnect.com

MERCHANT NAVY

Merchant Navy Welfare Board

30 Palmerston Road
Southampton
Hampshire, SO14 1LL

Tel.: 023 8033 7799
Website: http://www.mnwb.org/
E-mail: enquiries@mnwb.org.uk

Royal Alfred Seafarers' Society

Head Office Weston Acres,
Woodmansterne Lane,
Banstead, Surrey, SM7 3HA

Tel.: 01737 353763
Website: http://www.royalalfredseafarers.com/
E-mail: royalalfred@btopenworld.com

Shipwrecked Mariners' Society

1 North Pallant,
Chichester,
West Sussex, PO19 1TL

Tel.: 01243 789329
Website: http://www.shipwreckedmariners.org.uk/
E-mail: grants@shipwreckedmariners.org.uk

Trinity House

Corporation of Trinity House,
Tower Hill,
London, EC3N 4DH

Tel.: 020 7481 6900
Website: http://www.trinityhouse.co.uk/

ARMY

The Guards Associations (Grenadier, Coldstream, Irish, Scots, Welsh)

Regimental Headquarters
Wellington Barracks
Birdcage Walk
London, SW1E 6HQ

Tel. 020 7414 3280

Army Catering Corps Association

ACC Association,
Dettingen House,
Princess Royal Barracks
Deepcut,
Surrey GU16 6RW

Tel.: 01252 833394
E-mail: accassociation@rhqtherlc.org.uk
Website: http://www.accassociation.org/

Enquiries regarding benevolence grants or welfare should be directed to:
The Controller Benevolence
RHQ The RLC
Dettingen House
Deepcut
Surrey, GU16 6RW

Airborne Forces Security Fund

RHQ PARA Merville Barracks
Colchester,
Essex, CO2 7UT

Tel.: Civil (01206) 81 7078 or 01206 817079
Military (9) 46660 7078
E-mail: syfund@parachute-regiment.com

Army Benevolent Fund

Mountbarrow House,
6-20 Elizabeth Street,
London, SW1W 9RB

Tel.: 0845 241 4820
Website: http://www.armybenfund.org/index2.html
E-mail: enquiries@armybenfund.org

ATS & WRAC Association Benevolent Fund

Gould House
Worthy Down
Winchester
Hampshire, SO21 2RG

Tel.: 01962 887612
Website: http://www.wracassociation.co.uk/BenFund.html
E-mail: benfund@google-mail.com

Royal Artillery Charitable Fund

Artillery House
Royal Artillery Barrack
Larkhill
Salisbury
Wiltshire, SP4 8QT

Tel.: 01980 845895

Royal Army Educational Corps Association

RAEC Association
Worthy Down
Winchester
Hampshire, SO21 2RG

Tel.: 01962 887870
E-mail: raec,association@milnet.uk.net

The Royal Army Ordnance Corps Charitable Trust

Dettingden House
Princess Royal Barracks
Deepcut

Camberly
Surrey, GU16 6RW

Tel.: 01252 833376
E-mail: raocassociation@rhqtherlc.org.uk

Royal Army Pay Corps Regimental Association

RHQ AGC
AGC Centre
Worthy Down
Winchester
Hampshire, SO21 2RG

Tel.: 01962 887436
Website: http://www.rapc.co.uk/
E-mail: regsec@rapc.co.uk

Royal Army Service Corps and Royal Corps of Transport Benevolent Fund

Dettingen House
The Princess Royal Barracks
Deepcut
Camberley
Surrey, GU16 6RW

Tel.: 01252 340898

Royal Engineers Association

RHQ Royal Engineers
Brompton Barracks
Chatham
Kent, ME4 4UG

Tel.: 01634 847005
Website: http://www.reahq.org.uk/
E-mail: info@reahq.org.uk

Riflemen's Aid Society

RHQ
The Royal Green Jackets Peninsula Barracks
Romsey Road

Winchester
Hampshire, SO23 8TS

Tel.: 01962 828526
E-mail:rfnaid@royalgreenjackets.co.uk

Royal Signals Association

RHQ Royal Signals
Blandford Camp
Dorset, DT11 8RH

Tel.: 01258 482090
E-mail: rsa@rsignals.net

Royal Tank Regiment Benevolent Fund

RHQ RTR
Stanley Barracks
Bovington
Wareham
Dorset, BH20 6JA

Tel.: 01929 403331
Website: http://www.royaltankregiment.com/
E-mail: regtlsec@royaltankregiment.org

Special Air Service Regimental Association

SAS Regimental Association Office
The Occupier
PO Box 35051
London, NW1 4WF

Website: http://www.marsandminerva.co.uk

ROYAL AIR FORCE

Princess Mary's Royal Air Force Nursing Service Association

COS Health / DGMS (RAF)
Nimrod Block
CHQ Wy
RAF High Wycombe

Walters Ash
High Wycombe, HP14 4UE

E-mail: association@pmrafns.org
Website: http://www.pmrafns.org

Royal Air Force Association

RAFA Central Headquarters,
117 Loughborough Road,
Leicester, LE4 5ND

Tel.: 0116 266 5224
Fax: 0116 266 5012
Website: http://www.rafa.org.uk

Royal Air Force Benevolent Fund

67 Portland Place,
London, W1B 1AR,

Free support line: 0800 169 2942
Website: http://www.rafbf.org/

The Royal Air Forces Ex-POW Association

Secretary, John K. Banfield
Flat 1
Lincoln Lodge
21 Downs Court Road
Beckenham, Kent BR3 6LW

EX-SERVICE ORGANISATIONS

The Association of Jewish Ex-Service Men and Women

Shield House
Harmony Way
Off Victoria Road
London, NW4 2BZ

Head Office, Tel. 020 8202 2323
E-mail: headoffice@ajex.org.uk
Website: http://www.ajex.org.uk/

Royal Commonwealth Ex-Services League

Haig House
199 Borough High Street
London, SE1 1AA

Tel.: 020 3207 2413
E-mail: http://www.commonwealthveterans.org.uk/

The British Korean Veterans Association

For further information contact:
Frank Ellison OBE, BEM, JP
Hon. Secretary
12 Fields Crescent
Hollingworth
Hyde
Cheshire, SK14 8RJ

E-mail: frank.ellison@btinternet.com

British Limbless Ex-Servicemen's Association – BLESMA

185–187 High Road
Chadwell Heath
Romford
Essex, RM6 6NA

Tel.: 0208 590 1124
Fax: 0208 599 2932
E-mail: headquarters@blesma.org
Website: http://www.blesma.org/

The Burma Star Association

4 Lower Belgrave Street
London, SW1W 0LA

Tel.: 020 7823 4273
Website: http://www.burmastar.org.uk/

Canadian Veterans Associations of the UK

Contact: P.G. Mercer Esq.
National Secretary
Canadian High Commission

MacDonald House
Grosvenor Square
London, W1X OAB

Tel.: 0171 629 9492

Chindits Old Comrades Association

National Headquarters
TA Centre
Wolseley House
Fallings Park
Wolverhampton
West Midlands, WV10 9QR

Website: http://www.chindits.info/COCA/Main.htm

Dunkirk Veterans Association

Contact: J.E. Horton
White Wickets
Parkgate Road
Neston
South Wirral
Cheshshire, L64 6QQ

Tel.: 0151 3363530

Far East (Prisoners of War and Civilian Internees) Fund

30 Copsewood Way
Bearsted
Maidstone
Kent, ME15 8PL

Tel.: 01622 737124

Gulf War Veterans Association – Office Closed

No contact info on site other than e-mail address:
e-mail larry@gulfveteransassociation.co.uk
Website: http://www.gulfveteransassociation.co.uk/
Last newsletter 2005

Normandy Veterans Association

Contact: E.S. Hannath Esq.
General Secretary
53 Normandy Road
Cleethorpes
South Humberside, DN35 9JE

Tel.: 01472 600867

Officers' Association

1st Floor Mountbarrow House
6–20 Elizabeth Street
London, SW1W 9RB

Tel.: 0845 873 7153
E-Mail: postmaster@officersassociation.org.uk
Website: http://www.officersassociation.com

Polish Ex-Combatants Association

240 King Street
London, W6 0RF

Tel.: 020 8741 1911

Royal Star & Garter Homes

Richmond Hill (CC)
Richmond
Surrey, TW10 6RR

Tel.: 020 8439 8000
E-Mail: generalenquiries@starandgarter.org
Website: http://www.starandgarter.org

Royal Hospital Chelsea

Royal Hospital Road
London, SW3 4SR

Website: http://www.chelsea-pensioners.co.uk/home
Tel.: 020 7881 5200

The Sir Oswald Stoll Foundation
446 Fulham Road
London, SW6 1DT

Tel.: 020 7385 2110
Fax: 020 7381 7484
E-mail: info@oswaldstoll.org.uk
Website: http://www.oswaldstoll.org.uk/

Haig Homes – Housing for the Ex-Service Community
Alban Dobson House (CC)
Green Lane
Morden
Surrey, SM4 5NS

Tel.: 020 8685 5777
E-mail: haig@haighomes.org.uk
Website: http://www.haighomes.org.uk

St Dunstan's – An Independent Future for Blind Ex-Service Men and Women
12–14 Harcourt Street (CCWeb)
London, W1H 4HD

Tel.: 020 7723 5021
E-Mail: enquiries@st-dunstans.org.uk
Website: http://www.st-dunstans.org.uk
St Dunstan's has two established centres: one in Brighton and one in Sheffield. Another is due to open in Llandudno shortly.

CIVILIAN AND HOUSING CHARITIES

Abbeyfield Society
Abbeyfield House
53 Victoria Street
St Albans
Herts, AL1 3UW

Tel.: 01727 857536
E-mail: enquiries@abbeyfield.com
Website: http://www.abbeyfield.com

Anchor Trust
Customer Contact Centre
Milestone Place
100 Bolton Road
Bradford, BD1 4DH

Tel.: 0845 140 2020
Website: anchor.org.uk

The Salvation Army
UK Headquarters
101 Newington Causeway
London, SE1 6BN

Tel.: 020 7367 4802
E-Mail: margaret.meldrum@salvationarmy.org.uk
Website: http://www.salvationarmy.org.uk

Salvation Army Social Work Great Britain and Ireland

101 Newington Causeway
London, SE1 6BN

Tel.: 020 7367 4879
Website: http://www.salvationarmy.org.uk

Young Men's Christian Association

640 Forest Road
London, E17 3DZ

Tel.: 020 8520 5599
E-Mail: enquiries@ymca.org.uk
Website: http://www.ymca.org.uk

St Christopher's Hospice

51–59 Lawrie Park Road (CC)
Sydenham
London, SE26 6DZ

Tel.: 020 8768 4500
E-mail: info@stchristophers.org.uk
Website: http://www.stchristophers.org.uk/

The Mental After Care Association (MACA)
1st Floor
Lincoln House
296–302 High Holborn
London, WC1V 7JH

Tel.: 020 7061 3400
Fax: 020 7061 3401
E-mail:info@together-uk.org
Website: http://www.together-uk.org

Leonard Cheshire Disability
Head Office
Leonard Cheshire Disability
66 South Lambeth Road
London, SW8 1RL

Tel.: 020 3242 0200
Fax: 020 3242 0250
E-mail: info@LCDisability.org
Website: http://www.lcdisability.org/

Sue Ryder Care
114–118 Southampton Row
London, WC1B 5AA
Telephone: 0845 050 1953
E-mail: info@suerydercare.
Website: http://www.suerydercare.org/

Richmond Fellowship
80 Holloway Road (CC)
Highbury & Islington
London, N7 8JG

Tel.: 020 7697 3300
E-mail: communications@richmondfellowship.org
Website: http://www.richmondfellowship.org.uk/

CIVILIAN AND MEDICAL CHARITIES

Age Concern England
Room CC09 Astral House
1268 London Road
London, SW16 4ER

Free helpline 0800 00 99 66
E-Mail: customer.relations@acent.co.uk
Website: http://www.ageconcern.org.uk

Age Concern and Help the Aged in Scotland
Causewayside House (SCC)
160 Causewayside
Edinburgh, EH9 1PR

Tel.: 0845 833 0200
E-Mail: enquiries@acscot.org.uk
Website: http://www.ageconcernscotland.org.uk

Alcoholics Anonymous
PO Box 1
10 Toft Green
York
North Yorkshire, YO1 7NJ

Tel.: 01904 644026
Website: http://www.alcoholics-anonymous.org.uk

Alcoholics Anonymous (Scotland)
Northern Service Office (SCC)
Suite 442 Baltic Chambers
50 Wellington Street
Glasgow, G2 6HJ, Scotland

Tel.: 0141 226 2214

Alzheimer's Society
Devon House
58 St Katharine's Way
London, E1W 1JX

Tel.: 020 7423 3500
E-Mail: enquiries@alzheimers.org.uk
Website: http://www.alzheimers.org.uk

Diabetes UK

Macleod House
10 Parkway
London, NW1 7AA

Tel.: 020 7424 1000
E-mail: info@diabetes.org.uk
Website: http://www.diabetes.org.uk/

Epilepsy Action

New Anstey House
Gate Way Drive
Yeadon
Leeds
West Yorkshire, LS19 7XY

Tel.: 0113 210 8800
E-Mail: epilepsy@epilepsy.org.uk
Website: http://www.epilepsy.org.uk

The British Psychological Society

St Andrews House
48 Princess Road East
Leicester LE1 7DR

Tel.: +44 (0)116 254 9568
Website: http://www.bps.org.uk/home-page.cfm
E-mail: enquiries@bps.org.uk

British Red Cross

44 Moorfields
London, EC2Y 9AL

Tel.: 0844 412 2804
E-Mail: information@redcross.org.uk
Website: http://www.redcross.org.uk/

British Tinnitus Association

Ground Floor
Unit 5 Acorn Business Park
Woodseats Close
Sheffield
South Yorkshire, S8 0TB

Telephone: 0800 018 0527 free of charge – from within the UK
only 0845 4500 321 local rate – from within the UK only
Website: http://www.tinnitus.org.uk/
E-mail: info@tinnitus.org.uk

The Stroke Association

Stroke House (CC)
240 City Road
London, EC1V 2PR

Tel.: 020 7566 0300
E-Mail: info@stroke.org.uk
Website: http://www.stroke.org.uk

Crossroads – Caring for Carers

10 Regent Place
Rugby
Warwickshire, CV21 2PN

Tel.: 0845 450 0350
Website: http://www.crossroads.org.uk/

Disabled Living Foundation

380–384 Harrow Road (CC)
London, W9 2HU

Tel.: 0845 130 9177 or 020 7289 6111
Website: http://www.dlf.org.uk/
E-mail: info@dlf.org.uk

Family Action (Formerly Family Welfare Association)
Family Action Central Office
501–505 Kingsland Road
London, E8 4AU

Tel.: 020 7254 6251

Grants Service
Family Action
501–505 Kingsland Road
London E8 4AU

Tel.: 020 7241 7459 (2 p.m. to 4 p.m. on Tuesday, Wednesday
and Thursday only)
Website: http://www.family-action.org.uk

Elizabeth Finn Care (Previously DGAA Homelife [Distressed
and Gentlefolk's Aid Association])
1 Derry Street
London, W8 5HY

Freephone: 0800 413 220 (24 hours)
E-mail: info@elizabethfinn.org.uk
Website: http://www.elizabethfinncare.org.uk
Tel.: 020 7396 6700 (office hours)

**The Firefighters Charity (Formerly the Fire Services
Benevolent Fund)**
The Fire Fighters Charity
Level 6, Belvedere
Basing View
Basingstoke
Hampshire, RG21 4HG

Tel.: 01256 366566
E-mail: info@firefighterscharity.org.uk
Website: http://www.firefighterscharity.org.uk/

Help the Aged

Head Office
207–221 Pentonville Road
London, N1 9UZ

Tel.: 020 7278 1114
E-mail: info@helptheaged.org.uk
Website: http://www.helptheaged.org.uk

Huntington's Disease Association

Neurosupport Centre
Norton Street
Liverpool
Merseyside, L3 8LR

Tel.: 0151 298 3298
E-mail: info@hda.org.uk
Website: http://www.hda.org.uk/

MENCAP (Royal MENCAP Society)

Mencap National Centre
123 Golden Lane
London, EC1Y 0RT

Tel.: 020 7454 0454
Fax: 020 7608 3254

E-mail: mailto:information@mencap.org.uk
Website: http://www.mencap.org.uk

The Mental Health Foundation

9th Floor Sea Containers House
20 Upper Ground
London, SE1 9QB

Tel.: 020 7803 1100
E-Mail mhf@mhf.org.uk
Website: http://www.mentalhealth.org.uk

Mind (National Association for Mental Health)

Granta House

15–19 Broadway
London, E15 4BQ

Tel.: 020 8519 2122
E-mail: contact@mind.org.uk
Website: http://www.mind.org.uk/

Multiple Sclerosis Society

MS National Centre
372 Edgware Road
London, NW2 6ND

Tel.: 020 8438 0700
Fax: 020 8438 0701
Website: http//:www.mssociety.org.uk

Scotland – National Office
Ratho Park
88 Glasgow Road
Ratho Station
Newbridge, EH28 8PP

Tel.: 0131 335 4050
Fax: 0131 335 4051

Wales – National Office
Multiple Sclerosis Society Wales/Cymru
Temple Court
Cathedral Road
Cardiff, CF11 9HA

Tel.: 029 2078 6676
Fax: 029 2078 6677

Northern Ireland – National Office
The Resource Centre
34 Annadale Avenue
Belfast, BT7 3JJ

Tel.: 02890 802 802
National Website: http://www.mssociety.org.uk/

Parkinson's Disease Society of the UK

215 Vauxhall Bridge Road
London, SW1V 1EJ

Tel.: 020 7931 8080
E-Mail: enquiries@parkinsons.org.uk
Website: http://www.parkinsons.org.uk

Queen Elizabeth's Foundation

Leatherhead Court
Woodlands Road
Leatherhead
Surrey, KT22 0BN

Tel.: 01372 841100
E-mail: info@qef.org.uk
Website: http://www.qefd.org

Royal National Institute of Blind People (RNIB)

105 Judd Street
London, WC1H 9NE

Tel.: 020 7388 1266
Fax: 020 7388 2034
E-mail:helpline@rnib.org.uk
Website: http://www.rnib.org.uk

Royal National Institute for Deaf People (RNID)

Head Office
19–23 Featherstone Street
London EC1Y 8SL

Tel.: 020 7296 8000
E-mail:informationline@rnid.org.uk
Website: http://www.rnid.org.uk/

Samaritans

The Upper Mill (356LEG)
Kingston Road
Ewell
Surrey, KT17 2AF

Tel.: 08457 909090
E-mail:jo@samaritans.org
Website: http://www.samaritans.org/
Write: Chris, P.O.Box 9090, Stirling, FK8 2SA

St. Joseph's Centre
Holy Cross Hospital
Hindhead Road
Haslemere
Surrey, GU27 1NQ

Tel.: 01428-656-517

info@sagb.co.uk
http://www.sagb.co.uk/

Victim Support
Hallam House
56–60 Hallam Street
London, W1W 6JL

Tel.: 020 7268 0200
Website: http://www.victimsupport.org.uk/

Index